SIDDHARTH SAREEN

T0284048

THE SUN ALSO RISES IN PORTUGAL

Ambitions of Just Solar Energy Transitions

BRISTOL
UNIVERSITY
PRESS

First published in Great Britain in 2024 by

Bristol University Press
University of Bristol
1–9 Old Park Hill
Bristol
BS2 8BB
UK
t: +44 (0)117 374 6645
e: bup-info@bristol.ac.uk

Details of international sales and distribution partners are available at
bristoluniversitypress.co.uk

British Library Cataloguing in Publication Data
A catalogue record for this book is available from the British Library

ISBN 978-1-5292-4210-2 paperback
ISBN 978-1-5292-4211-9 ePub
ISBN 978-1-5292-4212-6 ePdf

Cover design: blu inc
Front cover image: alamy/Liubomir Paut-Fluerasu

For Annie, Indus and Fado
My trusty companions on solar energy governance
fieldwork in Portugal and so much more.
Thank you for allowing me to give this book the
time that it took to research and write!
This labour of love is dedicated to you, with love
from Siddharth (Papa)

***

And to the memory of Gieve, who gave of himself so
generously and had so much to give: you never got to
read this book, but I have read you, and through that
you are in my writing.

Contents

List of Figures and Table

About the Author

Siddharth Sareen, born 1988, is a human geographer, development researcher and political ecologist. He works on the governance of energy transitions at multiple scales. He has researched and taught on these themes in a wide range of countries and academic institutions from 2011 onwards. He completed this book in 2024 when working as Professor of Energy and Environment at the Department of Media and Social Sciences, University of Stavanger, and Professor II at the Department of Geography and the Centre for Climate and Energy Transformation, University of Bergen. He lives in Norway with his wife, daughter and dog.

Acknowledgements

The author acknowledges the University of Stavanger and the University of Bergen for their support in making this research and book possible. The author also thanks the Research Council of Norway for funding the Accountable Solar Energy TransitionS (ASSET) project (grant 314022), the Trond Mohn Foundation and the Akademia Agreement for enabling early parts of the research, the European Union's Horizon 2020 for funding time on Eurosolar4All (grant 101032239) and the University of Bergen for funding the open access publication of this monograph. This work would not have been possible without the hundreds of interlocutors who took the time and trouble to generously share their perspectives and valuable insights, and for this the author owes them a debt of gratitude that this book marks a modest effort to repay.

ONE

Introduction

Urgency, justice and scales in the Portuguese solar energy transition

Lisbon City Hall is a historic building that stands proud and majestic at Praça do Município, a stone's throw from the touristy and sprawling grand square of Praça do Comércio in downtown Lisbon. On 21 August 2018, I had an audience with one of the city's top representatives in this imposing building, accompanied by one of the city's foremost solar experts working for Lisboa Enova, the municipal energy agency. She and I shared reflections on our way down in the elevator after the meeting, where we had discussed Lisbon's solar energy ambitions at length. I congratulated her on Lisbon having won the accolade of the European Green Capital 2020, and asked what it meant for the city. She replied wryly that the biggest thing the award achieved for the city was to speed up the signing of files from the lower floors to the upper floors of the building we were in. At the time, I remember being struck by the pragmatism of her statement. It was a mark of intent, much like Portugal's approach to the solar energy transition it had begun to embrace, which had drawn me to Lisbon. Figure 1.1 shows the layered street landscape typical of Lisbon.

Figure 1.1: Layered street landscape typical of Lisbon

A year later, at the Urban Future Global Conference in Oslo in May 2019, I was fortunate to witness the passing of the torch between the two cities. Oslo was the European Green Capital for 2019, and at the closing event of the conference at a posh hotel in central Oslo, the team representing Lisbon took the stage, my interlocutor from the previous summer among them. The Lisbon team had big plans, a packed calendar full of events for their award year of 2020. As luck would have it, Lisbon and Portugal – like so much of the rest of the world – closed down most activities from March 2020 onwards. As I spent two-and-a-half calendar years away from the country, I had a chance to reflect upon a follow-up interview with the same interlocutor on 31 July 2019, my last one in Portugal until I could next visit in February 2022.

I asked how it felt to visit Oslo during its European Green Capital award year, and if it had inspired their thinking about their plans in Lisbon. She mused that they had expected the streets to be filled with activities representing the championing of sustainability, and that they were hard at work to make that vision come true throughout 2020 in Lisbon. How tragic, then, that the majority of those planned events never came to pass, or had to play out in a society where the streets were relatively empty during the COVID-19 pandemic. Yet the city's vision of sustainability has continued, strengthened by Portuguese encounters with wildfires and floods that catapulted sustainability onto the political agenda in a big way, making the 2019 national election perhaps the first 'climate election' in Portugal (Fernandes and Magalhães, 2020).

★★★

To appreciate the sea change that was underway in Portugal when I visited Lisbon City Hall back in 2018, it is worth recounting some major political developments from the first quarter of the 21st century. Historically, political leadership has swayed between Socialist Party and Social Democratic Party led governments since the constitution of the Assembly of the Republic in 1975 (Magone, 2014). This was the year following the Carnation Revolution of 25 April 1974 that saw a peaceful military coup end four decades of fascist rule under the authoritarian Estado Novo regime. The Socialist Party was in power for two periods during 2005–2011, with a weakened but adequate mandate in the election of 2009. These were difficult years to govern Portugal, as the country faced significant economic hardship during the global economic recession from 2008 onwards (Magalhães, 2017). This overlapped with some poorly drawn-up wind energy contracts for large concessions awarded by the government to Energias de Portugal, the formerly state-owned energy utility that completed progressive privatization in 2011. As wind energy costs fell during the 2000s, the company was in possession of state contracts awarding it lucrative fixed tariffs, and had an incentive to delay commissioning. This meant windfall profits from wind energy development when the plants did come online, built at considerably lower cost while continuing to be rewarded with highly subsidized tariffs for years to come (Peña et al, 2017). The resultant furore combined with the economic downturn meant anti-incumbency politics peaked, and in the

2011 national election, the Social Democratic Party came to power. While this major political shift can be largely attributed to the troubled national economy, the prevalent mood meant that the missteps on energy looked especially bad.

This ushered in a period of reduced renewable energy subsidies. Attractive household rooftop solar energy tariffs that had led to rapid growth in the diffusion of small-scale solar deployment were reduced to a meaningless scheme, and uptake died down by 2013. This is emblematic of how the government during this period did not differentiate between one form of renewable energy or another, or between different forms of ownership or the spatial scales of deployment. As a solar energy researcher who used to manage a photovoltaics testing lab told me on 27 September 2017 in Evora, renewable energy had been 'weaponized' for the 2011 elections, and this stalled any progress in the years that followed. So when the Socialist Party assumed power again after the 2015 national elections, the prospects looked none too rosy for the future of solar energy in this country with the highest solar irradiation rate in Europe.

This was the puzzle that drew me to study solar energy development in Portugal from 2017 onwards. Why, I wondered, was there so little of it in a country so abundant in this free resource and so lacking in fossil fuels, for which it relied on expensive imports? Moreover, given Portugal's reputation as a leader in wind energy – something that it did manage albeit at great cost to a depleted exchequer – why would it not develop solar energy as a complementary clean energy source, to secure a high share of renewable energy throughout the year? Surface explanations tend to point to the Portuguese cultural trait of being risk averse (Tejada, 2003). Richer economies like Germany had invested considerably in the 'energiewende' or energy transition, and in the aftermath of the 2008–2015 recession, Portugal had little appetite to invest in large infrastructural changes when it was cutting already-low public sector salaries through a particularly bleak form of austerity politics. Moreover, large neighbour Spain had introduced a so-called 'sun tax' in 2015 thanks to its conservative Popular Party, leading to years of uncertainty and reduced economic returns for households investing in rooftop solar modules, in the name of protectionism for electricity utilities. The risk-averse Portuguese did not want to play with fire when they were already struggling to make ends meet, or so the argument goes.

Yet, I was only partly convinced by such a rationale. Portugal had, after all, been host to what was the world's largest solar plant when it was being built in 2008 – the 36 megawatt (MW) Amareleja plant in Moura, in the Alentejo region of solar-rich southern Portugal. The truth seemed to be that Portugal lacked state capital as well as political capital to invest in subsidizing renewable energy. The Socialist Party had been burned in its early tryst with wind energy, and in 2017, the political messaging was clear: no subsidies for solar energy.

★★★

Globally, however, the energy sector was in flux, and Portugal was no stranger to these trends. Solar energy costs had plummeted throughout the 2010s, reaching parity with other energy sources at a rate few had imagined possible as recently as 2000. Consequently, solar energy plants were becoming attractive investments, and countries were launching reverse e-auctions to determine the competitive market price for concessions. In May 2017, solar auctions in India's Rajasthan created a stir by setting a world-record low tariff during bidding for Bhadla solar park – the world's largest – at below €30 per megawatt-hour (MWh), breaching parity with coal thermal plants (Bose and Sarkar, 2019).

The Secretary of State for Energy in Portugal settled upon an approach where Portugal would champion solar energy, but without subsidies, based on this being the cheapest and most competitive energy source. Already in July 2017, there were announcements of the first solar plant being constructed, backed by Irish WElink Energy and its strategic partner China Triumph

International Engineering, and this 46 MW 'Ourika' plant commenced operation in June 2018, as the largest unsubsidized solar plant on the Iberian peninsula. This was subsequently acquired by Allianz Capital Partners, a leading investor in this space, through a 20-year fixed price power purchase agreement. This success and the global trend led the Portuguese government to roll out its own solar reverse e-auctions in July 2019. The newly founded Ministry of Environment and Energy Transition (hereafter MATE) specified a ceiling tariff of €45 per MWh, considerably cheaper than the annual average cost of about €55 per MWh on the Iberian wholesale electricity market Mercado Ibérico de Electricidade (MIBEL).

What followed is historic. The first Portuguese solar auction brought in an average tariff of just above €20 per MWh, less than half the specified price cap, which was already significantly lower than the annual average price on MIBEL. The lowest bids for auction lots even went down to €15 per MWh, leading to concerns of overly thin profit margins and feasibility. It is one thing for solar auctions to set a world record for the world's largest solar park – Bhadla is sized at a mammoth 2,245 MW, and India was already well on its way to becoming a relatively mature and sizeable solar energy producer by 2017. It is quite another matter for a small southwestern European country, with well below a gigawatt (GW) of installed solar capacity at the time, to break this world record, clearing 1.15 GW of auctioned lots, with 250 MW settled shortly thereafter. Nor was this a one-hit wonder. In August 2020, Portugal successfully ran its second solar auctions of 670 MW despite the COVID-19 pandemic, with one of the bids setting another world record at €11.14 per MWh.

Solar was certainly here to stay, even if only time would tell the feasibility of commissioning solar plants with the prospect of the lowest tariffs secured. Indeed, given the supply chain uncertainties and construction delays brought about due to pandemic circumstances, timelines for commissioning were relaxed from the originally specified two-year window in the auction tenders, and the plants from these auctions were allowed grace periods while retaining eligibility for the tariffs secured in auction. A cautionary deposit (on a fixed percentage basis, this amounts to a large amount given the scope of these solar projects in the tens and even hundreds of MW) ensures that bids are earnest.

***

There is a story to be told about policy development in parallel. As a Member State of the European Union (EU), Portugal released its National Energy and Climate Plan 2030 in 2019. This aimed to expand solar energy to over 9 GW by 2027. This was a tremendously ambitious commitment, which at the time represented an approximately tenfold increase in installed solar capacity within eight years. Compared to a total installed capacity of approximately 17 GW for electricity in Portugal, this also meant embracing a major ambition to develop large amounts of solar power. Solar plants run at peak output for only limited parts of the day and year, with a relatively reliable output profile that exhibits considerable variation with a midday peak and naturally zero production after sunset until sunrise. Thus, aiming to increase solar capacity to this large extent meant also embarking upon a transition to a totally different energy system. Sure enough, this is what the country embraced in its Carbon Neutrality Roadmap 2050, which was promoted simultaneously, with a progressive vision of electrifying and decarbonizing many sectors of the national economy, while exponentially increasing renewable energy generation (Gil and Bernardo, 2020).

In 2021, Portugal became one of the over-achievers of the Powering Past Coal Alliance, when it shut down the last of its coal thermal plants in Sines, a mammoth 1.2 GW power plant south of Lisbon, thus exiting coal nine years ahead of its 2030 target (Blondeel et al, 2020). This left major electricity transmission infrastructure available to stream clean energy generation from this area. Consequently, many companies explored prospects to develop solar plants in its proximity. Spanish developer Iberdrola received a go-ahead to build a 1.2 GW solar plant in Santiago do Cacém southeast of the closed coal plant, symbolically named Fernando Pessoa after the celebrated national poet. In addition to clearing a plantation of 1.5 million eucalyptus trees, this solar development involves felling 1,800 cork oaks. These are heavily protected in traditional Portuguese landscapes, and linked with enormous cultural and economic value. The bark of these trees is harvested in nine-year cycles, with Portugal home to a third of the world's cork oaks and half the global production of cork. These cultural landscapes, shaped through human activity, are called 'montados'. They constitute multifunctional agro-silvo-pastoral ecosystems, where cork is harvested from cork oaks older than 25 years, that can live for several hundred years. The Portuguese Environment Agency allowed razing 1,800 of these trees for the solar plant, which would be the largest in Europe and is expected to be commissioned in 2025. Figure 1.2 shows cork oaks in the Alentejo region.

Alongside this momentous development in 2023, Portugal released its new National Energy and Climate Plan 2030, which now targets 20.4 GW of installed solar capacity by 2030.

Figure 1.2: Cork oaks in the Alentejo region

This is more than double the previous – already ambitious – target, and would represent an eightfold increase from the approximately 2.5 GW of solar energy plants installed by 2023. These developments not only make this book timely, but give room for pause, considering that this book maps my front row seat to Portugal's expanding solar ambition from 2017 to 2023. During a five-year period within these seven years, Portugal increased its installed solar capacity fivefold, from approximately half a GW at the end of 2017, to over 2.5 GW by the end of 2022. Much of this was low-hanging fruit, built where there was existing electricity transmission capacity, and where land was readily available to lease on a long-term basis. Projections by Bloomberg New Energy Finance suggest that from 2023 onwards, Portugal will add approximately 1.6 GW solar annually until 2026, reaching 9 GW by then.[1] This would leave a need to more than double capacity during 2026–2030.

To look at this another way: by 2023, Portugal has crossed 50 per cent of its electricity production coming from renewable energy sources (Russo et al, 2023). In the new plan, it now aims to increase this to 80 per cent by 2026, as part of its plan for a net zero economy by 2045. There is no dearth of ambition. Of the 20.4 GW Portugal has set its sights on, 14.9 GW is to come from utility-scale plants. However, more than a quarter – 5.5 GW – is to be sourced from more spatially distributed generation. If 2017–2023 marks the coming-of-age story of Portuguese utility-scale solar, which looks set to continue along the arc that it is traversing, then to achieve this crucial quarter of the target, 2023–2030 will need to see community-scale solar blossom. Progress has been painfully slow in this respect, but hearteningly, community-scale solar does not raise the same ecosystem trade-offs as utility-scale solar projects are so often bound up with. The challenges it has faced are largely bureaucratic and technocratic, and the general trend across EU Member States in the 2020s shows improvements on this.

The spatial scales of a just solar transition

The story of a just transition to low-carbon systems has become a well-rehearsed one. It has been increasingly written about by researchers and reporters since the 2010s, and the role of solar energy has come to be recognized as prominent (McCauley et al, 2023). The fact that solar energy has been the fastest growing source of energy in the world in the 2020s makes it particularly important what role it plays in enabling a just transition. Considerable research has made it evident that a chief advantage of solar photovoltaics, or PV – in terms of the sociospatial impact – is its modular nature. Solar panels can be installed as a few kilowatts (kW) or tens of kW on a household rooftop, a few hundred kW on say a public school roof or in a field as ground-mounted solar modules, or as a few MW in a more rural area. This makes it possible for solar plants to fulfil energy demand in a highly localized way, from serving individual household needs to feeding nearby industrial plants. With the advent of legislation on energy communities in an increasing number of (especially European) countries, it is becoming feasible for the same community-scale solar plant (often in the hundreds of kW) to generate electricity for multiple users within a radius of several kilometres (Hewitt et al, 2019). This is virtual consumption, in the sense that the electricity feeds into an electricity distribution grid that is connected beyond this radius and carries more content than simply the production from this solar plant, but its usage is territorially bounded, making location part of the eligibility conditions to be recognized as an energy community in most emerging regulations.

Thus, whereas most solar capacity addition to date has come from large, utility-scale solar plants in the tens, hundreds, or even thousands of MW, there are exciting examples of other forms of spatial emergence and proliferation, as well as other models of social organization and ownership of these emergent solar energy infrastructures. The city of Amsterdam, for instance, has supported solar installations by households and local organizations by undertaking large procurements of solar modules in order to play the intermediary role of quality assurance and ethical procurement, while enabling access to economies of scale for its residents through these bulk purchases. This has led to the installation of hundreds of MW within the city boundaries, with more growth expected in years to come, an impressive and very desirable achievement in terms of sociospatial justice effects. Meanwhile, for less savoury reasons in South Africa – which has faced a power crunch and massive load shedding due to complications involving its electricity utility ESKOM – some 3.2 GW of distributed solar energy capacity was built within 18 months during 2021–2023, a remarkable development. This is larger growth than the world's largest solar park, within a short timeframe, is owned by a large number of entities and spatially spread across the country rather than being concentrated at a single – typically remote – location. Such patterns usually mean a more equitable distribution of benefits, as a wider set of owners experience the gains that flow from harnessing widely available solar energy and lowering their reliance on a fickle and expensive electricity grid. These patterns also invariably lead to greater energy system efficiency, as energy production comes up close to energy demand, reducing the need to develop resource-intensive electricity transmission infrastructure (which has significant associated carbon emissions), and lowering the technical and commercial energy losses linked to transmitting and distributing electricity over distance.

★★★

There are people aplenty who have championed the promise of community-scale solar within Portugal's energy transition. My own research has often placed this in the foreground as a desirable priority since 2017, and in the course of conducting over a hundred interviews and spending extended periods of fieldwork in Portugal, I have encountered many actors pushing for rapid change along these lines. Yet until 2023, the story was largely one of frustration: with very sluggish movement, with tokenism, and with the massive scale of the challenge to be overcome for community-scale solar even as utility-scale solar began to make quite rapid progress. As the first formal actor in this regard, a solar energy cooperative called Coopérnico was heavily involved in both public policy discussions and backstage pushes to promote movement for community-scale solar projects (Wittmayer et al, 2022). In the early 2020s, others emerged, including a grassroots solar energy community in the Telheiras parish of Lisbon, and one led by the energy agency of Almada municipality in a social housing cooperative. This latter was part of a project called Eurosolar4All that I was involved in during 2021–2024, funded through the European Commission's Horizon 2020 programme. Other initiatives cropped up, including in conjunction with research projects, or floated by companies eager to find ways to increase off-grid local electricity consumption, for purposes similar to a power purchase agreement with solar developers operating a solar plant close to their energy-intensive operations. Perhaps most notably, companies such as CleanWatts and GreenVolt took up roles as intermediaries, becoming commercial players who could facilitate a rapidly replicable techno-economic model for energy communities by including hundreds of households in each, notably targeting smaller Portuguese towns.

From early 2021 onwards, all of these initiatives were spurred on by high retail electricity tariffs, as the Russian war in Ukraine and the design of electricity markets in Europe (which sets the reference price based on the price of the last unit sold to meet demand on the wholesale electricity market) sent the price of grid electricity spiralling up and left many household economies reeling. This is also when Portugal's Recovery and Resilience Plan kicked in, allocating 13.9 billion Euro in grants and 2.7 billion Euro in loans from the EU into the national economy during 2021–2026, with 38 per cent of this amount supporting climate objectives and another 22 per cent fostering the digital transition. This put in place funding to support long-sighted interventions, for instance to enhance energy efficiency and support demand response in Portugal's future energy system (Ribeiro, 2022). In the meantime, with European Commission approval in June 2022, Portugal was able to control electricity prices, channeling 2.1 billion Euro in subsidies to power producers, largely for electricity based on fossil fuels like gas and coal. In September 2022, the government also announced 2.4 billion Euro in a 'Families First' programme to support low-income households against the high cost of energy. These targeted fiscal policies were stop-gap measures that highlighted the societal cost of not having a boom in small-scale solar plants. The performance of the first of CleanWatts' renewable energy communities, which began to function in August 2021 in Mirando do Douro, put this in perspective by recording savings of over €30,000 for its members in one year.[2] By August 2022, the company had established a hundred communities with over 17 MW and 1,500 members with a similar model, and envisaged adding tens of MW and thousands of members a year across Portugal.

Yet it took until summer 2023 for licences from the Directorate General of Energy and Geology (hereafter DGEG) – the national executive agency for Portugal's electricity sector – to come through for the small Telheiras renewable energy community in Lisbon and the one serving the social housing cooperative in Almada, despite each involving small local rooftop solar installations of just some tens of kW. Musing on these small and hard-to-come-by successes compared to the rapid strides Portugal had made in utility-scale solar installations in the early 2020s, I was taken back to some words that had stayed with me from an interview at the MATE on 7 March 2019. The well-intended governmental advisors I met with that day were astute in their plan to "move aggressively on large-scale solar", which they explained meant attracting large foreign capital investment to build massive solar capacity of several GW that they lacked the wherewithal for within the national budget. As the saying goes, it is expensive to be poor, and Portugal's credit rating after the economic recession of 2008–2015 made for a high cost of capital on loans, not allowing for state-financed development of renewable energy along a rapid timeline. The advisors were also strategic in their outlook on community-scale solar energy. One said:

'We debate these issues about community energy, but across Europe they are still mostly pilot level. We also have a political timeline going up to elections. We are doing homework, most likely for the next administration to implement. Baby steps but we want to make the right ones. We want to allow all possibilities, not to close doors.'

Frustratingly, four years later, it was indeed pilots that Portugal was witnessing when it came to community-scale solar, pilots that had already been years in the making and were yet to reach an operational stage. As one of the advisors argued back in March 2019:

'We are not in the futurology business here. The grid as a supporting structure has to be financed. So hyper-decentralized structures are long-run scenarios. Timeline to elections is too short. There are legal obstacles on small-scale, so on that we are trying to strike that down to open up options.'

Yet does treading gingerly and proceeding slowly by a government in fact make it a participant in 'the futurology business', condemning community-scale solar energy to a slow rate of emergence in a rapidly evolving solar sector, despite its social desirability?

★★★

Hindsight is 20/20, so I am careful not to cast accusations, but rather to appreciate the historically contingent nature of particular decisions and related outcomes. It is in such instances that long-term ethnographic fieldwork offers a lens unmatched for its revelatory power towards an understanding of energy politics and the evolving political economy of a sector so vital to society, so bound up with the national economy, and so often pivotal to the changing tides of political fortune. The coalition led by the Socialist Party did indeed win the national election on 6 October 2019, which stayed in government, so this calculated approach can be said to have paid rich dividends, with its position on solar energy and energy transitions more generally a crucial piece of the political puzzle. The government, through consultation with the DGEG and the Energy Services Regulatory Authority and formally through the Assembly of the Republic (national Parliament), followed through on this position voiced during our meeting, issuing Decree-Law No. 162/2019 on 25 October 2019. This reorganized the legal regime applicable to the self-consumption of renewable energy. It left strategic room for flexibility in the evolution of related rules, to receive feedback from sectoral stakeholders and evolve regulation adaptively.

The next major legislative move came more than three years down the line. This time, it was a fortnight *before* snap elections, which had been declared on 4 November 2021 by the incumbent Socialist Party, to be held on 30 January 2022. On 14 January 2022, Parliament passed Decree-Law No. 15/2022, which provided for the regulation of collective self-consumption, and repealed its predecessor. This legislation addresses the commercial relations between entities involved in Renewable Energy Communities (RECs); establishes the legal bounds for data measurement, reading and availability; specifies the basis for modes of sharing energy between self-consumers; and governs the applicability of regulated tariffs to RECs.

While I was on fieldwork in Portugal during February–March 2022, I found limited appreciation of this adaptive approach to developing REC regulations – dating back to 2019 – among stakeholders. On 4 March 2022, an expert in a municipal energy agency that was moving quite aggressively on establishing a solar energy community reflected that: "With the new REC regulation of late 2019, and the amendments until January 2022, planning as a frontrunner is hard ... we are spending tens of thousands of Euro on legal analysis to ensure that the contracts [with member households] are compatible with REC regulations." Along similar lines, an expert in solar energy research and development (R&D) for a multinational energy company explained: "In R&D our roadmap is more like five years, so we are interested in developing with a vision to have a better product more suited to REC definition, and then pass it on once developed to [the commercial arm of the company] for upscaling." These perspectives capture the seasoned take of solar sector practitioners that the regulations prevalent by 2022 did not constitute an inviting prospect for commercial or public operators, but rather had a broad signalling effect for solar energy actors to develop techno-economic models to enable the rollout of solar energy communities in the mid- to late 2020s in ways well-suited to the overarching sectoral developments by then. This is somewhat contrasted by the emergence and growth of players such as CleanWatts, yet installing some tens of MW of community-scale solar by 2023, while impressive, is hardly a bedazzling accomplishment compared to over 2.5 GW of large-scale solar plants.

It seems unfortunate that a country with high solar ambitions and apparent political stability and momentum, led by the Socialist Party since 2015 until at least 2026 after the 2022 elections, made such little headway until 2023 in terms of a just solar energy transition. Compared with developments across Europe and worldwide, however, this is in keeping with the general trend of large-scale solar plants dominating deployment, and large multinational companies owning most emergent solar energy infrastructure (Sareen and Haarstad, 2021). Despite all its modular characteristics and promises for spatially distributed and socially more just models of ownership and control, to a large extent solar energy remains of a piece with the historical form of the energy sector from its fossil fuel era: dominated by players with access to large financial capital, who command preferential access to decision-makers and benefit from favourable regulations, and exercise top-down control over the means of energy production and the ability to corner profit from cheap resources, even encompassing free sunlight. Surprisingly, across the Atlantic Ocean, Portugal's large Lusophone cousin Brazil exhibits a counter-trend (Rigo et al, 2022), with 21 GW of its 30 GW of installed solar capacity by mid-2023 coming from spatially distributed solar systems, of which 10 GW from household PV systems below 7 kW each.[3] Of course, Brazil is nine times larger than Portugal, which explains its far larger installed capacity. But it also has 20 times the population of Portugal, meaning its population density is more than twice as high, thus evidently it is not easier land availability that drives this difference in patterns. Without getting into other social justice issues, where Brazil faces its own complex challenges, this stark difference is a reminder that another form of solar energy transitions is possible, one more attuned to sociospatially just outcomes.

This brings us to the question at the core of this book, in a country aiming to surpass 20 GW of solar by 2030 with over a quarter of it as small-scale solar plants: will the sun rise to such great heights in Portugal, and in doing so, will it take people along in just ways?

TWO

Methodology

An energy researcher's tryst with Portugal during 2017–2023

It was past five in the afternoon on a sunny winter day on 27 February 2019 in placidly beautiful Faro, capital of the southernmost Portuguese region of Algarve, when the then-Secretary of State for Energy took to the bright stage of the dimly lit Auditorium 1.4 of the University of the Algarve. A large turnout in this ample hall had sat for over two hours, taking in what the various speakers brought together by the Ministry of Environment and Energy Transition (MATE) had to say about the Roadmap for Carbon Neutrality 2050 (RNC2050) and the 2019 version of the National Energy and Climate Plan 2030. To the credit of MATE, they were taking the recently launched RNC2050 on a roadshow to the north and south of Portugal, bringing the eminent speakers assembled here out of their comfortable quarters in Lisbon, to reach a wider and varied audience. Disappointingly but unsurprisingly, the ones who came to listen were largely men, rather typical of the energy sector in general. I was the only one in the room who did not speak fluent Portuguese – in fact, I spoke none at all, and could primarily follow the content from projected slides, thanks to figures and some understanding of the written language. A live audio translation tool that I tried to run from my phone did not function very well from Portuguese to English, though it did help with the odd bits and bobs. I enjoyed the Secretary's animated talk. Unlike some others, he did not read from a script, and was clearly a confident and effective orator. He emphasized that the energy transition was an economically sound, necessary decision for the government to have taken. He talked it up as a timely and great opportunity, and a strategic priority for Portugal. Then he indicated two parallel processes that needed to interact, and as an indicator of some critical things I missed over the two-and-a-half hours that afternoon (and on some similar occasions), I could not quite catch what these were, though I surmised that they had to do with steering a strong energy market while decarbonizing the Portuguese economy.

I was sitting a row behind the Secretary and had taken the opportunity to pass him a policy brief from the Centre for Climate and Energy Transformation in Bergen – based on our research on solar energy transitions in Portugal and other related challenges – at the start of proceedings. He had accepted this with a smile and glanced at it. So when I made the bold move to raise a hand and pose a question about the process of developing community-scale solar after his talk, he had some sense of who I was. He responded: "[B]ottom-up is very important and so is decentralized [renewable energy development] and we will do just that." He also addressed a question by an industrial innovation panellist on the need to unlock the potential of community energy for the people. He continued to explain that at the same time, there is a role for strong centralized production, and the question of grid capacity poses a trade-off. He talked about building up an energy system that is able to capitalize on the promise of multiple things simultaneously, like digitalization, cost decreases in renewables, and the emerging potential role of aggregators as Portugal moves towards a flexibility market, all of which requires enabling legislation, while addressing the historical sectoral debt and dealing with the existing grid capacity configuration. He flagged the role of key sectoral players like the Directorate General of Energy and Geology (DGEG) and the Portuguese Environment Agency in enabling solar growth and resolving grid capacity questions in line with due process. He also highlighted the need for municipalities to participate in energy transitions, and closed by underscoring the generally positive response to the intent of energy transition.

I knew that my question had gone down well, because one of the most important people in the room that day and in the Portuguese energy sector in general, took the time to show up at a meeting I had scheduled at the National Laboratory for Energy and Geology the following week. He spoke with me for over two hours, and later sent me a link to a job opening with him, encouraging me to move to Portugal to work on their energy transition. I was flattered but politely declined, already living my dreams working as a postdoctoral researcher on the governance of energy transitions with a base at the University of Bergen along the beautiful western coast of Norway with its flat hierarchy, which I enjoyed in work environments. But two things from this set of interactions nonetheless stayed with me: a sense of assurance that my research was striking some relevant chords in the 'real world', and curiosity as to whether there weren't already heaps of smart and driven people in Portugal trying to enable its energy transition.

That was relatively early on in my tryst with the Portuguese solar energy transition: at the time I had spent about three months in the country over two years of running a research project to understand this fascinating unfolding phenomenon. In the years since, I have become convinced that I do have something relevant to offer through my research on these developments, both to the Portuguese people and to others elsewhere interested in the promise of just solar energy transitions. This has led to numerous research and policy outputs, from journal articles and book chapters to policy briefs and magazine articles, as well as many a meeting, seminar, training school, academic workshop and public talk. I have also had the good fortune to meet many of the dedicated and talented people contributing to turning the ambitious vision of a solar transition into reality in Portugal, and doing significant work to ensure that this happens with salutary effects for society. Yet in conducting over a hundred interviews (108 formal ones to be accurate, but there are so many additional informal discussions that feed into my understanding that specifying a number does not really do them justice; even so, my interview notes exceed 100,000 words), I have become acutely conscious of some privileges that allow me to make a uniquely valuable contribution. Hence my decision to write this book, to dig a little deeper and to spread my insights a bit wider. Towards that, in this chapter, I offer some details on how I have gone about investigating and addressing the central question of this book, replete with details about fieldwork techniques, and reflections on limits in scope.

★★★

I first encountered the enigma of Portugal's solar energy transition – or to be more precise the relative lack of it – when working on a research consultancy for the Regulatory Assistance Project alongside my first postdoctoral fellowship, based between New Delhi in India and Erfurt in Germany, during 2016–2017. The research consultancy was part of a 15-state comparative project called 'Mapping Power' on the political economy of electricity distribution in India, as part of which I studied these trends in Rajasthan and Gujarat. I was keen to take up a medium-term postdoctoral project on the political economy of solar energy from 2017 onwards, because the global trend of cost declines and grid parity had become apparent to any curious observer, and it seemed likely that solar energy would soon become a bigger part of the energy sector in most countries with high solar irradiation. I had very limited experience with energy sector governance, having conducted my doctoral research at the

University of Copenhagen during 2012–2016 on indigenous people's access to forests, extractive industries, and the nature of authority in regions with resource conflicts. A three-year position would allow me to come into my own as a human geographer, political ecologist, and development researcher working in the emergent interdisciplinary field of energy social science during the mid- to late 2010s. As a country facing quandaries over its own energy transition, Norway was early in recognizing the need for such research, and a national endowment had helped establish the centre in Bergen which posted a postdoctoral position. When conducting fieldwork for the research consultancy in 2016, I was captivated by the developments I had seen on solar energy in Rajasthan, and struck by the lack of similar movement in Europe's resource-rich regions. Researchers were already studying Spain, the most prominent case, and some of the southeastern European economies were still coping with the difficult aftermath of the economic recession and unlikely to pull together an ambitious plan for solar energy transition. Yet I was fascinated by prospects for solar in Portugal – a country I had never been to at the time – a young, stable democracy having exited the authoritarian Estado Novo regime as recently as 1974, nestled next to its much larger Iberian peninsula neighbour Spain in the relatively isolated, unperturbed southwestern corner of Europe. I wondered: if the sun also rises in Portugal, how will ambitions of a just solar energy transition fare?

Curiously, almost no energy social scientists seemed to have been struck by similar concerns, or at least to have acted upon them sufficiently to publish in-depth research, unlike in the heavily studied contexts of Germany, California and other early movers. Surely it was interesting and relevant to understand how more financially constrained countries could enable rapid and just solar energy transitions? So I wrote a proposal, successfully competed for the position, and set up shop in Bergen, a corner of northwestern Europe some 2,600 kilometres from Lisbon as the crow flies, and more like

3,700 kilometres by surface transport, and got to work on this fascinating problem from April 2017 onwards.

And the rest, as they say, is history. Or at least it is a part of my publishing record during 2018–2023 that I take pride in. Table 2.1 shows 26 journal articles and book chapters published by October 2023 that draw on my research on solar energy transitions. Twenty are sole-authored or led by me, six by others as part of their doctoral dissertation or Master thesis work, or within combined empirical analysis across cases drawn from mine and others' work (1, 2, 9, 12, 14, 17). Most draw directly on my solar energy research in Portugal, while some work along a tangent from it, either as introductions to edited volumes or special issues of journals that emerged from this focus over time (13, 20, 22), or as agenda-setting syntheses (10, 19) or thematic stocktakes (3, 15). Not all are equally relevant to this book, but it is important to highlight here that these were all made possible due to the sort of long-term ethnographic engagement I have had the privilege of nurturing with Portugal during this consequential period for its peculiar solar energy transition.[1]

What the listed research outputs do not capture is my evolution as a person and researcher. I spent about half my working time during 2017–2023 focusing on this theme, close to 7,000 hours. Spread over seven visits, ranging from six weeks to a single week, and usually three to four weeks each, I spent about 4,000 hours actually living in Portugal – half a year in total out of this seven-year period, despite the two-and-a-half year COVID-19-induced gap. Of this time, I spent about a fifth working, or nearly 1,000 hours, of which perhaps 200 were devoted to actually conducting interviews. The rest of the time went towards dozens of field visits to solar plants and related sites of energy infrastructure and decision-making, to moving between places in Lisbon and elsewhere in Portugal, to soaking in observations and spending time taking notes and reflecting upon analytical insights. I also opened countless tabs on web browsers, taking in detailed information on specific sectoral developments, parsing thousands of articles in

Table 2.1: Siddharth Sareen's research outputs during 2017–2023 based on the study of solar energy transitions in Portugal

1   Silva, L. and Sareen, S., 2023. The calm before the storm? The making of a lithium frontier in transitioning Portugal. *The Extractive Industries and Society*, 15, 101308. https://doi.org/10.1016/j.exis.2023.101308

2   Scharnigg, R. and Sareen, S., 2023. Accountability implications for intermediaries in upscaling: Energy community rollouts in Portugal. *Technological Forecasting and Social Change*, 197, 122911. https://doi.org/10.1016/j.techfore.2023.122911

3   Sareen, S., Sorman, A.H., Stock, R., Mahoney, K. and Girard, B., 2023. Solidaric solarities: Governance principles for transforming solar power relations. *Progress in Environmental Geography*, 2(3), pp 143–165. https://doi.org/10.1177/27539687231190656

4   Sareen, S., Shokrgozar, S., Neven-Scharnigg, R., Girard, B., Martin, A. and Wolf, S.A., 2023. Accountable solar energy transitions in financially constrained contexts. In B. Edmondson (ed) *Sustainability Transformations, Social Transitions and Environmental Accountabilities* (pp 141–166). Cham: Springer International Publishing. http://doi.org/10.1007/978-3-031-18268-6_6

5   Sareen, S., Grandin, J. and Haarstad, H., 2022. Multiscalar practices of fossil fuel displacement. *Annals of the American Association of Geographers*, 112(3), pp 808–818. https://doi.org/10.1080/24694452.2021.2000850

6   Sareen, S., 2023. Solar spectacles: Why Lisbon's solar projects matter for energy transformation. In H. Haarstad, J. Grandin, K. Kjærås and E. Johnson (eds) *Haste: The Slow Politics of Climate Urgency* (pp 234–242). London: UCL Press. https://www.uclpress.co.uk/products/194544

7   Sareen, S., 2022. Legitimating power: Solar energy rollout, sustainability metrics and transition politics. *Environment and Planning E: Nature and Space*, 5(3), pp 1014–1034. https://doi.org/10.1177/25148486211024903

8   Sareen, S., 2022. Drivers of scalar biases: Environmental justice and the Portuguese solar photovoltaic rollout. *Environmental Justice*, 15(2), pp 98–107. https://doi.org/10.1089/env.2021.0048

9   Mahoney, K., Gouveia, J.P., Lopes, R. and Sareen, S., 2022. Clean, green and the unseen: The CompeSA framework: Assessing competing sustainability agendas in carbon neutrality policy pathways. *Global Transitions*, 4, pp 45–57. https://doi.org/10.1016/j.glt.2022.10.004

Table 2.1: Siddharth Sareen's research outputs during 2017–2023
based on the study of solar energy transitions in Portugal (continued)

10	Sareen, S. and Wolf, S.A., 2021. Accountability and sustainability transitions. *Ecological Economics*, 185, 107056. https://doi.org/10.1016/j.ecolecon.2021.107056
11	Sareen, S., 2021. Digitalisation and social inclusion in multi-scalar smart energy transitions. *Energy Research & Social Science*, 81, 102251. https://doi.org/10.1016/j.erss.2021.102251
12	Silva, L. and Sareen, S., 2021. Solar photovoltaic energy infrastructures, land use and sociocultural context in Portugal. *Local Environment*, 26(3), pp 347–363. https://doi.org/10.1080/13549839.2020.1837091
13	Sareen, S., 2021. Energy infrastructure transitions and environmental governance. *Local Environment*, 26(3), pp 323–328. https://doi.org/10.1080/13549839.2021.1901270
14	Geels, F.W., Sareen, S., Hook, A. and Sovacool, B.K., 2021. Navigating implementation dilemmas in technology-forcing policies: A comparative analysis of accelerated smart meter diffusion in the Netherlands, UK, Norway, and Portugal (2000–2019). *Research Policy*, 50(7), 104272. https://doi.org/10.1016/j.respol.2021.104272
15	Sareen, S. and Haarstad, H., 2021. Decision-making and scalar biases in solar photovoltaics roll-out. *Current Opinion in Environmental Sustainability*, 51, pp 24–29. https://doi.org/10.1016/j.cosust.2021.01.008
16	Sareen, S. and Nordholm, A.J., 2021. Sustainable development goal interactions for a just transition: Multi-scalar solar energy rollout in Portugal. *Energy Sources, Part B: Economics, Planning, and Policy*, 16(11–12), pp 1048–1063. https://doi.org/10.1080/15567249.2021.1922547
17	Nordholm, A. and Sareen, S., 2021. Scalar containment of energy justice and its democratic discontents: Solar power and energy poverty alleviation. *Frontiers in Sustainable Cities*, 3, 626683. https://doi.org/10.3389/frsc.2021.626683
18	Sareen, S., 2021. Scalar biases in solar photovoltaic uptake: Socio-materiality regulatory inertia and politics. In A. Kumar, J. Höffken and A. Pols (eds) *Dilemmas of Energy Transitions in the Global South: Balancing Urgency and Justice* (pp 78–92). Abingdon: Routledge. http://doi.org/10.4324/9780367486457-5

(continued)

Table 2.1: Siddharth Sareen's research outputs during 2017–2023 based on the study of solar energy transitions in Portugal (continued)

19	Sareen, S., Thomson, H., Herrero, S.T., Gouveia, J.P., Lippert, I. and Lis, A., 2020. European energy poverty metrics: Scales, prospects and limits. *Global Transitions*, 2, pp 26–36. https://doi.org/10.1016/j.glt.2020.01.003
20	Sareen, S. and Haarstad, H., 2020. Legitimacy and accountability in the governance of sustainable energy transitions. *Global Transitions*, 2, pp 47–50. https://doi.org/10.1016/j.glt.2020.02.001
21	Sareen, S. and Grandin, J., 2020. European green capitals: Branding, spatial dislocation or catalysts for change? *Geografiska Annaler: Series B, Human Geography*, 102(1), pp 101–117. https://doi.org/10.1080/04353684.2019.1667258
22	Sareen, S., 2020. Reframing energy transitions as resolving accountability crises. In S. Sareen (ed) *Enabling Sustainable Energy Transitions: Practices of Legitimation and Accountable Governance* (pp 3–14). London: Palgrave Macmillan. https://doi.org/10.1007/978-3-030-26891-6_1
23	Sareen, S., 2020. Metrics for an accountable energy transition? Legitimating the governance of solar uptake. *Geoforum*, 114, pp 30–39. https://doi.org/10.1016/j.geoforum.2020.05.018
24	Sareen, S., 2020. Social and technical differentiation in smart meter rollout: Embedded scalar biases in automating Norwegian and Portuguese energy infrastructure. *Humanities and Social Sciences Communications*, 7(1), pp 1–8. https://doi.org/10.1057/s41599-020-0519-z
25	Sareen, S., Baillie, D. and Kleinwächter, J., 2018. Transitions to future energy systems: Learning from a community test field. *Sustainability*, 10(12), 4513. https://doi.org/10.3390/su10124513
26	Sareen, S. and Haarstad, H., 2018. Bridging socio-technical and justice aspects of sustainable energy transitions. *Applied Energy*, 228, pp 624–632. https://doi.org/10.1016/j.apenergy.2018.06.104

Portuguese (often using an online translation tool) and English, and composing arguments by drafting text that I discussed with various friends and colleagues in rambling conversations and structured forums. But of the balance 3,000 or so hours, about

1,700 were spent not sleeping, but experiencing Portugal in its fullness and vibrance.

From the metropolitan bustle of Lisbon to the quiet expanse spread out around Evora, from the wondrous scenery of Sintra to the powerful waves of Nazare, from the historic landscape of the Douro valley to the enchanting festivity of Tavira, from the university town vibe of stony Coimbra to the medieval charm of Óbidos, from the cork oaks of Alentejo to the upbeat rhythms of Porto, from the glistening beaches of Vila Nova de Milfontes to the touristy trappings of Albufeira, and from the wooded hills of Monchique to the gleaming cobblestone streets of Santiago do Cacém, I came to know a country and its people better. Already invested in acquiring and improving Norwegian as a fourth language during this time, I did not make an effort to pick up Portuguese beyond the casual acquisition of some everyday vocabulary. My interviews handled sophisticated themes that required advanced language skills, and my interlocutors were typically – but not only – energy sector related workers who were well-versed in English, which made this our natural language of choice. Over many a *meia de leite* (my preferred coffee to order in Portugal), *pastéis de nata* (quintessentially Portuguese custard tart), *sumo de laranja* (orange juice), *croissant com queijo* (croissant with cheese, the go-to snack for a vegetarian) or *cerveja* (beer), I learnt a great deal about Portugal, the Portuguese and their solar energy transition.

Multi-sited and multi-scalar longitudinal ethnographic fieldwork

It is a rare privilege to have a long stretch of years to study the evolution of a sector in a context, especially when not primarily based in that context, and relatively early on in an academic career. I was fortunate to be able to focus primarily on Portugal's solar energy transition during 2017, while continuing to advance some completed energy research in India, and then during 2018–2019, split time between work on Portugal and on another project focused on smart electric meter diffusion in Norway. From 2020 onwards, I moved into a tenured academic position at the University of Stavanger, and brought in numerous externally financed research projects on the governance of energy transitions. I founded and led my own Sustainability Transformation programme area at the Faculty of Social Sciences, hiring several team members and working with a growing set of international collaborators, while continuing an additional engagement at the University of Bergen. From 2022 onwards, my performance led to a promotion to full professorship, all of which led to a great increase in time spent on project administration and various sorts of committee work. I taught and supervised postgraduate students throughout, as I think academics should.

Ordinarily, this might have meant that my interest in and research on Portugal fell by the wayside after 2020, considering that my growing portfolio of projects encompassed cases in over a dozen other countries.[2] By then, I had spent over four months in Portugal, splitting my fieldwork time between Lisbon and the rest of the country (primarily the Alentejo and Algarve regions) about equally. Yet a conceptual interest in accountable governance that matured through my research in Portugal led to a Research Council of Norway grant during 2021–2024, enabling me to continue and expand comparative research on solar energy transitions in Portugal and the Indian state of Rajasthan. This was both a chance to continue the explicitly multi-sited and multi-scalar empirical analysis I had been conducting since 2017, and also to consider how I could best utilize longitudinal ethnographic fieldwork.

Researchers are also people, and the single-most important event in my life in 2021 was the birth of my daughter. This meant my wife and I spent a year splitting parental leave between us, and that I set up fieldwork in Portugal in a manner compatible with bringing family along. This took place during February–March 2022, and we stayed mainly in Lisbon, with the protracted COVID-19 pandemic circumstances still weighing heavy upon people's daily lives. I was able to spend another span of close to a month in and around Lisbon during June–July 2023, this time with a toddler in tow, which meant considerable care work for my wife during her summer holiday. These personal circumstances are important to note, as they play roles in one's research: I was stricter about sticking to my working hours during these visits, and prioritized staying put in Lisbon with day trips out of town if required, but without the hassle of going around Portugal on overnight or longer trips as a family. We also brought along our Portuguese water dog, which further complicated travel, but was also delightful. My experience of Lisbon also changed, with more time spent in family rhythms in the evenings, visiting various parks in the city and engaging in child-centric leisure activities. I found a different window on to how solar energy manifests in urban environments, reflecting upon the same phenomenon from a new vantage point, much as someone gazing out at Lisbon from several of its *miradouros* (galleries with magnificent views over the city of seven hills) encounters it anew and works to piece together the city as a patchwork quilt of many complex and overlapping perspectives. Figure 2.1 shows rooftop solar in the city.

★★★

Visits to dozens of solar energy project sites, and to centres of expertise and activity in this sector around Portugal over the years, proved invaluable to enabling my continued probing of relevant themes even as the solar energy transition evolved. It became more clearly multiple transitions – not only the utility-scale plants that were now starting to proliferate with some rapidity in the early 2020s, but also small-scale solar plants whose rate picked up again, after stagnating since the early 2010s. The latter were driven by self-consumption solar plants put up by businesses, as they could meet a portion of electricity demand at rates significantly lower than industrial tariffs on the electricity grid. Households also took up solar

Figure 2.1: Rooftop solar in Lisbon

self-consumption, but not in a game-changing manner like that in Brazil. During interviews, I became interested in how stakeholders viewed the medium-term trends and prospects in this emergent trajectory. I returned to some key interlocutors in order to revisit themes from our discussions in light of developments. My own standing as a researcher with considerable outputs on the theme also opened up novel opportunities, for instance to organize a major European conference on Positive Energy Districts and give a plenary talk, and to participate in European project meetings with Portuguese public sector partners actually involved in implementing pro-poor community-scale solar plants based on policy learning from elsewhere.

While I have no crystal ball to gaze into the future, an empirical scientist can study contemporary trends to offer deep insights into a societal context. I met with all manner of sectoral stakeholders: from those working for solar energy cooperatives such as Coopérnico to those working with the large incumbent energy company Energias de Portugal (EDP); from those working within government agencies like the Energy Services Regulatory Authority (ERSE) and the DGEG to those working at research-cum-policy institutions like the National Laboratory for Energy and Geology and solar energy researchers at universities; from those with an activist interest in solar energy transitions to those working on dimensions of human change like members of the Tamera eco-community where I spent a fortnight back in 2017; from those participating as users in solar energy transitions to those making key decisions at ministries and investment banks; from those debating the social impact of energy transitions in European policies to those implementing measures in municipal energy agencies; from representatives of energy associations to solar installers and manufacturers; and from journalists shaping public discourse on and understanding of solar energy transitions to solar energy developers trying to carve out a place for themselves in a hyper-competitive and changeable sector.

To these diverse interviews and meetings, as well as to informal conversations, I brought an understanding that I would never be as experienced as most of these stakeholders in their particular domain of expertise, and certainly not in their relevant context-specific experience. To a senior expert within an energy sector organization, I would joke that "I'm not the one you call in to fix the wires!", but would also explain that I had the privilege of being paid to spend time discussing these issues with them and other sectoral stakeholders. This is a luxury seldom afforded to those with a deep working knowledge of any sector – they are busy and there is work to be done. It is in this way that I helped these valuable stakeholders make sense of my role as a researcher; they were important to aid my understanding and for their role in enabling (or not) Portuguese solar energy transitions. Additionally, my competence on governance issues in other transitioning sectors, based on research on urban transport transitions, digitalization, electricity distribution, urban planning and extractive industries, often allowed me to identify and articulate points of common interest. For instance, bringing insights from Norway's rapid transport electrification into a discussion about

Portuguese solar energy with an electricity regulator would lead to very fertile exchanges on smart grid development and implications for a future energy flexibility market. This in turn would produce insights that might come in handy during a site visit to a research and development facility of an energy company developing models for such an emerging market. And this understanding would be instrumental to discussions with a solar developer or an investment banker about the valuation of solar on the grid and prospects for competitive investments based on market evolution and the likely cost of capital given auction terms. Or an interview with an activist keen to enable more radical forms of solar energy communities would bring up resonance through my research with indigenous people fighting stacked odds to access forest resources for subsistence in a resource conflict region.

In all these interviews and conversations, I was forthright with my views, keeping an open mind while also probing where I was struck by some inconsistency, and being frank if asked for my take on the same subject or a related one. From the very outset, it was a priority for me to be able to talk freely with a wide variety of stakeholders. This meant conducting research on an anonymized institutional attribution basis, where I undertook to not identify respondents by name, nor usually even by exact institution, unless their statements were already public or clearly linked to a known position of a specific, unique institution. For instance, I mention 'solar developer' or 'energy journalist' without specifying a company or newspaper respectively, but do mention the Secretary of State for Energy or indicate that anonymized advisors represented a certain ministry. Similarly, I do not reveal identities of activists or particular staff in specific units of the executive agency DGEG or regulator ERSE, but do mention that someone represents EDP if the incumbent role is important to aid contextual understanding, or link a particular claim about energy cooperatives to Coopérnico given its distinct identity. In each of those cases, the person remains individually unidentified.

This is not only an important choice made with regard to thorough application of research ethics, but one that has been instrumental in building and maintaining trust within what is a highly networked sector that often deals with sensitive matters. That said, stakeholders with intimate knowledge of the sector may occasionally be able to surmise who made a particular statement; claims of this sort do not have a controversial nature, although some interlocutors were blunt in their views (and either comfortable with sharing them, or clear about saying something off the record). It is not uncommon for all the important decision-makers of the Portuguese energy sector – even beyond solar energy – to be assembled in one room. In a small country, the distance to power can be spatially small, and that can be reflected in information flow, even though power hierarchies and formal structures do exist in noticeable and consequential ways.

★★★

In many ways, then, this book is a palimpsest, which revisits seven years of productive research to glean something deeper, foreground something that has been out of focus. It takes critical pause to introspect, to reconsider, and perhaps to revise or to append a footnote to the analyses and claims that have gone before. There is a red thread that runs through the considerable set of outputs I have been involved in based on this research. I am deeply concerned with highlighting issues of relevance to urgency, justice, and for related reasons also the scalar aspects of the Portuguese solar energy transition. Like in the rest of my work, but more explicitly and with a bigger canvas for expression, this book addresses: what kind of spatial patterning is unfolding thus far and what are its sociopolitical implications for energy justice? Figure 2.2 shows an aerial view with some sense of an answer, as large solar and wind installations start to characterize changing Portuguese landscapes.

Figure 2.2: Aerial shot of large solar and wind plants

This is my way in to the challenge of answering whether Portugal will usher in a just solar energy transition. It is the same kind of 'futurology' that the advisors at MATE were unwittingly party to, except that I am consciously enrolled in it and committed to it. Moreover, I recognize the need for both urgency and justice as a call to action to be heeded far more acutely than the political realism of the advisors allowed them to do in 2019. This form of futurology is called 'prefigurative politics', meaning that I am part of the canvas that I am painting, and hold hope that my bold strokes can make a difference to the overall effect the painting produces. I want to create an effect that will move people to push for the future that, to become possible, needs us to proactively champion its cause. And who is the viewer of this solarpunk palimpsest? The viewer is constituted by multiple publics: policy makers, scholars, citizens with hopes and dreams of just solar energy transitions, solar developers, municipal workers. One way to find out whether a just solar energy transition is possible is to wait for 2030, but that is almost a guarantee that the answer will be 'no'. The powers that be will not make it come about of their own accord; they are too fogged up by structural protocols to sufficiently see through the smog of capitalist solar development towards a luminous citizen-centric solar future. Another way to answer the same question is to heed it as a clarion call: just solar energy transitions are calling, and to become possible they need each of us to step up and work towards them!

THREE

Solar Portugal 2017

From economic recession to unsubsidized solar energy projects in 2017

It was September 2017 when I first visited Portugal. Instead of heading to Lisbon, the national capital, I landed in Faro, the quaint capital of the Algarve region, the idyllic southern strip of Portugal that enthrals tourists and enriches residents with its sunny climate. After a fortnight living in the Tamera eco-community in rural Alentejo – a far less wealthy region than Algarve – I interviewed some experts in and around Faro. On 21 September 2017, this brought me to a researcher and administrator who had been entrusted with an important role in a national commission assessing a tragic forest fire from June 2017, which had led to the loss of dozens of lives in central Portugal. I have often thought back to this interview and the timing of my visit, as my departure from Portugal that October was followed by further tragic news of human deaths in northern Portugal due to a wildfire much later in the season than the *bombeiros* (volunteer firefighters) had expected. In this sense, my meeting with an important official in this process was bookended by two tragic occurrences: wildfires much earlier and much later than expected by the experts in charge of emergency preparedness (Silva et al, 2023), leading to significant casualties.

Portugal is no stranger to wildfires. Yet 2017 was an unusually terrible year in this respect, with some 21,000 wildfires in total, burning over half a million hectares of forest; and the two events of June and October caused the most grief and consternation, because between them they claimed 117 human lives, not to mention the loss of non-human nature.[1] These developments caused major consternation, and the Minister of the Interior stepped down from their post in October 2017 as a consequence. During the interview in September, I was struck by the fact that my interlocutor saw the energy sector as moving along as usual, without radical changes. He did not see the main university in the region, the University of the Algarve, as playing much of a role in national energy sector policies, nor see the region as being in a position to act on its strong solar prospects, with some of the highest irradiation in Europe. Coming from someone with an eminent position and a long history of engagement with territorial development related to the environment and other energy-adjacent sectors, this perspective revealed cause for concern. Clearly, the wildfires were manifestations of worsening climate change effects directly impacting Portugal through an extended wildfire season that stretched the limited capacity of their volunteer firefighter system. But there was no sense of urgent action to address the root cause of the problem, by accelerating climate change mitigation efforts in the vulnerable regions, and as concerningly, also a very limited ability to do much for requisite climate change adaptation. With stretched national budgets in a post-recession economy, forests were hardly getting huge room to make strategic investments in shifting forestry practices and vegetation; nor were these obvious measures to take when the problem was at a global scale, namely climate change. In Faro, the sense was that decisions and action had to stem from Lisbon.

***

This centralized national tendency in governance, and indeed energy governance, is typical of Portugal, as also in the energy sector more generally from a historical perspective. Energy has been a matter of national politics and policy, with big decisions made in the national capital, and regions following suit (Newell, 2021). To some extent in the early 21st century, cities have begun to mobilize to make their own climate change mitigation decisions, with climate budgets downscaled to the urban scale (Hsu et al, 2020). Portugal is no exception to these global trends. An analysis of its hydropower development convincingly demonstrates how this has been premised on internal colonial relations, with central state politics driving particular extractive patterns of development in vulnerable regions (Batel and Küpers, 2023). No doubt this energy infrastructure development has served Portugal as a whole well, providing domestic renewable energy resources to serve industrial demand in the northern parts of a country without its own base of fossil fuel sources and thus historically reliant on imports to meet part of its energy needs. But the benefits and burdens have been unequally spread across the landscape and population.

Following on from hydropower development in the late 20th century, research also details the rise of wind power in Portugal, following the example of Denmark (Bento and Fontes, 2015). Yet this development in the 2000s and 2010s did not lead to Portugal faring as well economically, due to poorly constructed wind energy contracts that imposed a large part of the cost on society while enabling windfall benefits for developers, notably the incumbent energy company Energias de Portugal through its renewable energy production arm. This led to considerable political intrigue and disenchantment with the energy transition (Silva and Pereira, 2019). In the actual places where wind energy infrastructure was situated, research shows considerable ambivalence in how this development was perceived by various societal actors both prior to and after development (Delicado et al, 2016). Solar energy is generally less disruptive than other energy sources to the surrounding landscape, but does have a considerable land footprint (Kiesecker and Naugle, 2017). Hence any plan to radically increase solar generation from the late 2010s onwards had a considerable knowledge base to draw on for strategic development, in order to intervene in society and landscapes in a manner that would be broadly acceptable, in addition to helping expand the Portuguese renewable energy sector to meet ambitious European Union targets as a Member State.

Portugal in 2017 was a riot of controversy. The minimum monthly wage was €557, barely enough to afford a hostel bed in a shared dormitory. By 2023, this had increased to €760, an increase of over one-third within six years, which is some indication of the burgeoning cost of living. Many workers I met in Lisbon lived an hour or more out of town to access affordable accommodation, spending hours on public transport to work in the city (also see Pereirinha and Pereira, 2023). This made for an uneasy juxtaposition with a different aspect of the same city – the rise of digital nomads sipping lattes at cafés along the beautiful cobblestone streets of the capital. The southwestern European capital, still a lot cheaper for tourists and upwardly mobile hipsters than larger European capitals like Paris or Rome, was becoming a huge drawing force for expats. In November 2016, it had hosted its first Web Summit, an arrangement that has continued since and that signals the vitality of the city and its intent to attract foreign capital.[2] But for many well-heeled expats, the draw of Portugal was not so much to its capital as to its controversial 'golden visa' programme, a residence permit for investment activity floated in 2012 as part of a multi-pronged strategy to claw its way back out of the economic recession. This offered perks like a low tax rate on global income for those willing to relocate to Portugal and invest in expensive property. Unsurprisingly, this led to considerable blowback from public sector workers facing pay cuts to their already meagre wages and pensioners being priced out of the homes they had expected to live out their days in, due to rising real estate value, lease rates and property taxes (Jover and Cocola-Gant, 2023). By 2023, the golden visa programme was being significantly adjusted in recognition of the righteous indignation of citizens. People were tired of an inflated housing market and the 'Airbnbization' of Lisbon's neighbourhoods (Sequera and Nofre, 2020).

Amidst all this, the Socialist Party-led coalition that came to power in 2015 had its work cut out to legitimate solar energy as part of its vision for a revitalized national economy that would benefit the Portuguese people. The devastating wildfire events of 2017 were no standalone accidents; they were precursors of more wildfires and floods to come. Back in 1755, the Great Lisbon earthquake had flattened the former city centre on a Saturday morning, ruining any planned All Saints' Day festivities on 1 November. The city had been subsequently rebuilt at a remove from its original centre, under the direction of the Marquês de Pombal, whose commemorative statue towers majestically over one of the city's major intersections, between the grand Avenida da Liberdade boulevard and the Eduardo VII Park. Figure 3.1 shows the Praça Marquês de Pombal. The political leadership was astute in its recognition of a need to combine climate change mitigation with broadly appealing elements in its political strategy, if it was to be re-elected in 2019 and stand a chance to define and implement a feasible agenda for energy transitions. When it came to solar energy, little had happened to move away from the stagnation that had set in since the early 2010s. So something had to be done, and soon. And it had to be done without imposing massive costs on an economy running on empty.

★★★

This is not how solar prospects always looked for Portugal. Back in the mid-2000s, it was host to the largest solar plant in development worldwide, namely Amareleja, planned as a 46 megawatt (MW) plant in Moura municipality of Beja district in the Alentejo region. By the time this was ready in 2008, it was no longer the world's largest, which is a testament to

Figure 3.1: Praça Marquês de Pombal

the rapidly growing nature of solar photovoltaics (PV) in the early 21st century with its spectacular cost declines through technological innovation and upscaled modular production led from China (Chase, 2019). A visionary mayor in rural Portugal dared to dream big, and attracted this investment by a Spanish major, Acciona Energy. A number of caveats were attached. The company would not only build the solar plant, it must also create 100 local jobs linked to a solar manufacturing facility, which would produce PV modules. Moreover, the project would invest in a PV panel testing facility, using state-of-the-art laboratory equipment to make Moura into a centre of competence for manufacture and quality assurance, in addition to high volumes of solar energy generation.

Ordinarily, this would have frightened most investors away, but the national government (also led by the Socialist Party during 2005–2009 and re-elected to serve during 2009–2011, before the Social Democratic Party came into power during 2011–2015) dangled a carrot that was simply too good to resist. At stake was a lucrative long-term feed-in tariff in the region of €300 per megawatt-hour, many times higher than the prevalent price on Mercado Ibérico de Electricidade, the wholesale electricity market for the Iberian peninsula.

Moreover, Acciona succeeded in acquiring the licence for what at the time was a massive amount of solar capacity, constituting about a third of the *total* solar capacity that could be licensed in Portugal in a year, which was capped at 150 MW at the time. So the costs it needed to pump into local solar sector development projects were to be weighed up in light of the spectacular profit margins it would be able to access, even given the higher cost of installing solar capacity back in 2008, which continued to come down manifold in the following years. All in all, having an enthusiastic political leader was far preferable to having bureaucrats who would create hurdles at every step along the way. There was the prospect of relatively simple land acquisition, and the availability of the electricity transmission grid within reasonable reach (32 kilometres), reducing the resource- and time-intensive process of acquiring permissions to build transmission infrastructure through land and then putting this in place.

Thus, the Amareleja solar plant came into being, and I had the good fortune to visit it along with a senior officer from Acciona in 2017, when it and the PV manufacturing facility had been up and running for nearly a decade. I will never forget the sense of awe I experienced upon seeing my first utility-scale solar plant up close. On 28 September 2017, my interlocutor drove me through the towering solar panels, and pointed out two hundred goats grazing within the fenced area. "Sometimes I feel like a farmer", he joked, also reflecting that "We've had more than 3,000 visitors, often groups, coming for a tour. Amaraleja is also a model of sorts to show what is possible with solar". I noticed that the capacity was specified as 36 MW and asked about this. His response was instructive:

'The 46 MW that was licensed eventually came up as 36 MW, because we ultimately used different panels with greater efficiency per unit area, and met the permitted cap on total generation (on an annual basis) with lower installed capacity than envisaged. We have been

producing the target number of hours consistently close
to 100 percent for 8–9 years now.'

This is yet more evidence of how rapidly PV technology
evolved, offering efficiency gains and reducing land
requirements. I was also struck by how high-tech the setup
was in general: panels were aerially monitored by drone, with
all their geolocations fed into a database, so that ones whose
trackers were malfunctioning could be realigned by a ground
team on a regular basis. Besides fencing, security took place
via remote surveillance. These aspects have become routine
for large solar plants, with Amareleja being a frontrunner. I did
reflect upon how this meant little in terms of local jobs linked
with the plant operation.

Over dinner, my interlocutor was kind enough to
introduce me to a Chinese counterpart who was managing
the PV module manufacturing facility, which Acciona had
subcontracted to Jinko Solar with its vast experience in this
sector. It was thus that I found myself in the rare position of
being given a behind-the-scenes tour of the PV factory, which
produced 25–40 MW of solar panels annually, depending on
what sort of panels were being made at what grade of quality.
Typical sizes were between half to one square metre of surface
area, for a peak output of 300 watts. My guide explained that:

'60 per cent of the cost is for the battery and silicon
screen, which cannot be imported from China to Europe
due to anti-dumping legislation, but are instead brought
from Taiwan for instance, where Chinese companies can
also run joint ventures. The workers had 12-hour shifts
earlier and protested, but now most have eight-hour shifts
and matters have been resolved smoothly.'

I was shown how the current assembly line was set up to make
panels with black frames instead of the more common grey
ones. Another insightful explanation followed:

'These cost more but are more aesthetic and preferred by some clients in the Nordic region who are willing to pay more, which is important because we are not in a position to compete on the market for the cheapest panels. Jinko Solar has expertise, but we are extremely small compared to the large setups in China which benefit from economies of scale.'

Two other interviews I had in Moura brought up this aspect of how tricky it was to navigate the solar sector economy – globally influenced by major Chinese production – from these rural backwaters of Europe. One was with a worker at the module testing laboratory, and the other with the iconic mayor himself, who was in his very last week in office.

At the Logica PV Lab, a worker showed me expensive equipment to test solar panels, and gestured around the crowded rooms:

'We have all these panel models that companies sent us to test. We conduct all the tests they need in order to get licences, but now our approvals have expired and we don't have the money to renew them. We have been running losses for three years, but it's because of bad management. They have no idea what these testing facilities are about or how to manage a project. They refused to give any co-funding so we couldn't compete for funds like most big competitors, and by the time we finally had orders coming in we couldn't afford our basic upkeep. I feel ashamed that things are in disarray like this.'

This was a startling illustration of the well-known theoretical proposition that it is hard to establish innovative value streams in a remote, rural region. Geography is not only spatial, it pertains to the embodied sense of a place, its culture. Here, the worker was explaining the lack of organizational wherewithal to use the initial infrastructural investment to carve out a niche in the accelerating PV industry. It was not simply that solar companies found it easier to ship panels to the large offices of TÜV in Germany for quality control and certification. Even the factory next door did not use this facility's local services. A special purpose vehicle floated by the municipality could not continue to make losses beyond three years, so the writing was on the wall: the laboratory would be shut down. "It is important that you see this", he continued, "the death of a vision, what was imagined and what happened".

I was able to meet the man behind this vision. Amareleja and the accompanying paraphernalia had come up under his watch, and then another member of his party had held office for successive terms. Thereafter, he had returned to office, and was about to end his term. The other member of his party was running again, but although we did not know it at the time of this interview, his party would lose, thus leading to power changing hands. He reflected on the regional development he had foreseen and played a major role in enabling:

'We floated a company and procured all the solar licences, then Acciona bought it for €17 million and spent another €3 million to set up community development funds. The solar park has come up and works for them but it does not have much to do with the local people. … The problem is that young people are all leaving places like Amareleja and Moura. Even my interest in regional history inspired my daughter to study Arabic and she has moved to an institute in Leiden in The Netherlands. Everybody who gets an opportunity in small places like this eventually uses it as a leg up to move to bigger cities. So we cannot be competitive with global industries.'

This was an incredibly perceptive and naturally politically savvy perspective. Unfortunately, it was also quite a gloomy one. My interlocutor went on to explicate a key conundrum for regional development even if there were to be an upcoming upswing for Portuguese solar: "Now there is no basis to get companies to invest in regional development initiatives because the government is not offering any incentives for solar, companies are only going to do it based on profit." His sage reflection was that with market mechanisms becoming the main driver of rollout, solar development would be uncoupled from its potentially wider societal functions, paving the way for commercially driven energy sector development. In reflecting upon the marketization of renewable energy, the pre-eminent theorist of rentier capitalism, Christophers (2022), sounded similarly concerned notes.

Soberingly, the Moura Solar Factory closed its doors just three months later, the day that Acciona had fulfilled the contract clause of providing 100 local jobs for ten years at the end of 2017. This was not the final nail in the solar coffin, however, as a new project revived prospects in 2021, with the company Lux Optimeyes Energy investing €5 million to manufacture flexible PV panels and lithium batteries from 2022 onwards, and aiming to employ 40 workers locally, primarily former Moura Solar Factory workers.[3] Perhaps competence developed in a region is not irrevocably lost, maybe it finds new ways forward. In all likelihood, the truth is somewhere in between, with small rural towns going through the challenge of a demographic shift with ageing residents, and youth struggling to find a foothold in the expensive and hyper-competitive cities of an increasingly neoliberal and dual-track economy. A just energy transition needs to address this, or it will face societal backlash or lead to polarization (Rodríguez-Pose, 2018). This is an economy where there are jobs and an attractive lifestyle for upper-middle-class workers and the uber-rich, but where those living on low public sector salaries or close to the humble minimum wage struggle to make ends meet in the absence of inherited family wealth. The growth of solar in Portugal has thus far meant little to this latter class.

The latest darling of the European solar scene

The story of the rise of solar in Portugal has become a familiar one to sectoral onlookers over the years. This is symbolized by prominent discursive shifts such as Lisbon's embrace of its identity as 'Lisboa Cidade Solar', or Lisbon Solar City, as a key aspect of its Sustainable Energy and Climate Action Plan. Defining such a plan was part of the municipality's partner status in the Covenant of Mayors for Climate and Energy, alongside over a thousand cities with climate goals. This included ramping up solar deployment within urban space, and electrifying numerous sectors like urban transport and waste management to decarbonize a range of economic activities along a defined, ambitious timeline. Since the early 2010s, Lisbon has been host to the Large Scale Solar Europe annual conference, and in 2023, it hosted the 40th edition of the International Photovoltaic Science and Engineering Conference, a long-running annual event since 1984 and the largest of its kind. This clear trend led me to label Lisbon 'the latest darling of the European solar scene' (Sareen, 2023, p 236).

It was not always like this.

In the heady 2020s, where solar is the world's largest growing energy source year after year, it is easy to forget the tensions and uncertainties of how the emergent sector looked in 2017. Building any sort of credible plan to accelerate solar deployment needed political gumption, but as the limited success of the Moura mayor's stratagem bears witness to, it also needed an economic fundament for continuity. After its debacle with wind energy financing, the Socialist Party could hardly afford to lose face in yet another renewable energy adventure. Moreover, solar was not the only game in town. In a meeting with a representative of the office of the Secretary of State for Energy on 3 October 2017, I was reminded of the complex, simultaneously technical and political nature of this puzzle:

'There is pressure to put up exploratory oil rigs due to examples like Norway. But that is a matter for DGEG [the Directorate General of Energy and Geology], and the Ministry, to decide on a technical basis. We are separate to guarantee that the Ministry operates autonomously and makes its technical decisions independently. The Secretary of State for Energy's office is a political one.'

Indeed, Galp Energia was pushing forward a consortium to prospect for offshore oil and gas west of Portugal. This was significant, as Portugal had used a lot of gas to meet demand in 2017, with hydropower faring poorly due to drought conditions. Historically, hydropower had grown in the 1970s, with large coal thermal plants coming up in the late 1980s and early 1990s. Portugal built up enough gas plants for strategic reserve capacity, and the Tâmega pumped hydro scheme was in the works (finally completed in 2022), enhancing options for energy flexibility. For DGEG as an executive agency, cost-effectiveness was a major concern, given an energy cost of €1.5 billion in 2015 and approximately €1.8 billion in 2017, with a total energy sector debt (tariff deficit) of about €5 billion in 2016 as per my interlocutor. A considerable part of annual cost came from the wind feed-in tariff legacy.

He continued:

'During 2012–13, energy costs were shown as low by extending an investor deal, due to end in 2020, until 2027. This placed a very heavy future public debt burden, but it was not presented that way. This is not easy to redress, because restructuring debt and changing the regulated tariff are linked. Debt is serviced through the grid access fee. But the regulated tariff is set as the price cap, as the tariff of last resort for the consumer. Right now, the feed-in tariff burden is phasing out too, and in a few years we will have fewer costs for consumers.'

Thus, from the government's perspective, they had inherited problems from a difficult past that made the present complicated to deal with. Grid access fees were high, and unpopular, but necessary to service high costs. This created pressure to keep variable costs low, meaning an incentive to chase after the cheapest energy sources.

Circling back to my interlocutor's observation about the institutional separation of the technical and political, it was apparent to me that they were in fact tightly linked in practice. The Secretary of State for Energy could hardly go out on a limb and promise a bright future for solar energy without checking with DGEG on how the sectoral economy looked, but equally, neither could DGEG simply proceed to make decisions on a technical basis if they would land the government in politically hot water. To make things even more complicated, a technical basis for decisions was based on far from perfect knowledge in a rapidly evolving energy sector. Indeed, renewable energy sources had breached cost parity with fossil fuels and thrown traditional calculations and assumptions in the air. One thing was certain, as my interlocutor emphasized: "As we change from a base load system to a flexible renewable energy based system, flexibility will have a return on investment."

What this implied for solar energy, with its predictable but variable output profile from morning through a midday peak until a late afternoon decline, was something only time and the configuration of a particular sociotechnical transition could tell. But it doubtless meant that I could not simply focus on PV, without also considering evolution in battery storage, smart grids, variable tariffs and other aspects linked to energy flexibility markets. For the sun to rise in Portugal in a way that cast its light far and wide, would entail a transformation of the whole energy sector.

★★★

With a jangle of thoughts in my head about what I had seen in Moura, trying to grasp the span of a decade's developments in the political economy of something as complex and as essential to society as the energy sector, I joined my interlocutor from Acciona Energy on his way into Lisbon. A month after my arrival in Faro, and beginning to make sense of Portugal and the solar adventure I was convinced it would embark on, I was about to head into the nerve centre of political decision-making, the national capital and fabled city of Lisbon.

As an Indian, I have mixed feelings for Portugal, a colonial power whose sweep of conquest included parts of India, most famously the western coastal state of Goa. Whereas Vasco da Gama is celebrated in his place of rest in the Jerónimos Monastery in the Belém parish of Lisbon near the Tagus river, my schooling taught me the ills of his conquests for the local ways of lives he invaded and destroyed. Yet in common across my research on energy in both these (and indeed numerous other) countries, I have a sense of great empathy with the people of Portugal. As a primer, I had read Barry Hatton's excellent introduction to the Portuguese (Hatton, 2012), tourist guides to Portugal (Rough Guides being an excellent pick) and a collection of the celebrated Fernando Pessoa's poems (Pessoa, 2006). As a vegetarian, I had eaten well in the vegan Tamera eco-community during my first fortnight in Alentejo, then been amused and charmed by the ability of many a small-town restaurant in Algarve to dish up a spinach lasagne (or else a pizza with strong *piri piri* upon request), after they had recovered from their initial consternation at my refusal to eat meat or their famous seafood.

More than differences, I found commonalities.

Yet nothing could have prepared me for the feeling of exhilaration when driving across the Ponte 25 de Abril, crossing south-to-north as we drove into Lisbon on a sunny late September day in 2017. Figure 3.2 shows an iconic view of the bridge across the Tagus river that links Almada to the south and Lisbon to the north. History connects us in countless ways, more than we will ever know, and I was entering the heart of a country that, until 1974, had suffered the imposition of an authoritarian regime. This was more

Figure 3.2: An iconic view of the Ponte 25 de Abril

recent than India's independence from British rule in 1947, and a more uniting part of the countries' shared history of struggle against oppression followed by periods of finding their feet and defining their role in the world. There was much in common, down to the recent history of unbundling energy sectors in the early 2000s (Ferreira et al, 2007), and gradually privatizing and liberalizing the energy sector. So Lisbon was new and familiar at the same time.

That is also true of the solar energy sector and how its key decision-makers doubtless felt when taking the first steps into the adventure that lay before them in 2017. With the whole country's energy future at stake, this delicately poised moment was a critical juncture, before everything exploded.

FOUR

Solar Portugal 2018–2019

A new ministry and a world-record solar auction in 2018–2019

When I entered a solar developer's office on the morning of 30 July 2019, the atmosphere was electric, with an air of palpable intensity combined with unprocessed disappointment. The bad news hadn't quite sunk in yet, it seemed so surreal. The interlocutor I was meeting was exhausted, and poured himself an espresso as he served me a cup. "I haven't slept in a week!", he exclaimed, before continuing more reflectively:

'We did not win any of the auctions, it has been a busy time but now there is a chance to relax. For developers like [us] this is not a good development, the lowest price going down to €14.70 per MWh [megawatt-hour]. For our classic business this is not good, it cuts our margins both in our management fees as well as EPC [engineering, procurement and construction] fees, as well as our premium on the greenfield development, which is our core business model. The business models have completely changed in these couple of weeks.'

These were raw reflections, before he had had a chance to do an internal wrap-up within the company. But already, he could see that the recent legislative changes that enabled the auctions allowed a promoter to develop specific grid connections to build their own additional projects. The auctions he was referring to were reverse e-auctions for solar projects with a specified amount of megawatts (MW) of availability at various locations on the transmission grid. The results had sent shockwaves around the world, with 1.15 gigawatts (GW) of the 1.4 GW capacity on offer being auctioned off at a highly competitive price, on average just above €20 per MWh, with the lowest bids sinking below €15 per MWh, as he mentioned. World records had tumbled, and the remaining 250 MW was allocated shortly thereafter at a similarly competitive price, all less than half the €45 per MWh ceiling tariff that the Portuguese government had set ahead of their highly anticipated first solar auctions.

The solar developer continued:

'We had a good [project financing] partner so we went pretty far, never imagined we would go down so low, since the partner could pay the premium and enable it. [Our] assumptions work with €17 per MWh, but if you cut assumptions further then you can make great margins with our existing projects, because they are pretty much ready to build and we have to do the EPC tender.'

So his main takeaway was that the auctions had served the role of revealing market fundamentals, and that costs for solar plants were even lower than previously calculated. Industry experts weighed in over subsequent days, noting that the lowest bids might well not turn out to be feasible, and that not all the auctioned capacity might result in actually installed solar capacity. But when investors put down money for projects in the hundreds of MW, it is evident that they covet the right to access limited grid capacity with licensed solar projects. In this sense, the government's gamble had delivered beyond its wildest dreams.

Figure 4.1: Large solar plant and transmission grid

The solar developer agreed. He voiced appreciation of the new framework, as one in tune with a sector with competitive costs, and a good way to avoid spending taxpayer money. He simply added that "for the auction there should be a floor, else the business becomes non-profitable, which is stupid as there is immense risk". This would require a lottery among willing bidders. In his view, a healthy project would need €26 per MWh, with an acceptable return on investment to justify the risk of a maturing sector. Figure 4.1 shows a large solar plant connected to the transmission grid. In general, he approved of the auction design, noting that the bid with the lowest costs was for a 150 MW project, which if not built, would attract a high penalty on the bank warranty, worth €60,000 per MW, so €9 million straight into state coffers. This talk of value brought us back to the matter of what the electricity grid itself was worth, at the lower scale of the distribution grid, with the expected move towards a flexibility market as solar and other renewable energy content increased. "It is hard to know the value of such a thing with all the future changes", said my interlocutor sagely.

He talked about the last auction lot, which had attracted a great deal of interest: "Now the municipalities all also understand that this is a good thing for them. Earlier we needed to convince them, now they have seen it work and they can see from others that it not only gives them money but also a good image." This was something that had changed – so much had changed! – since I last spoke with this solar developer on 23 October 2018. That was the month Portugal instituted a new ministry – the Ministry of Environment and Energy Transition – signalling its clear intent and priority to deliver on a bold new National Energy and Climate Plan 2030 (PNEC 2030), and its Roadmap for Carbon Neutrality 2050 (RNC2050). Back then, he had calmly listed four main criteria they assessed when planning solar projects: the distance to the grid; grid capacity availability at the nearest sub-station; land inclination (greater slopes can add solar installation costs and reduce generation); and land classification status, such as ZPE (protected zone), REN (reserved ecological) or RAN (agricultural) land. The cost difference was negligible compared to total project financing, which cost about €1 million per MW, but these criteria could affect processing time. Now, with the advent of solar auctions, developers could shoot for known grid capacity in specific locations, and have the benefit of a market mechanism to back their efforts when acquiring project financing. The promise of guaranteed revenue flow at a fixed tariff for 15 years, secured on a competitive basis through the auction design, eased financing and lowered the cost of capital.

In sum, from 2018 to 2019, what Portugal had witnessed – and succeeded in accomplishing – was the emergence of a competitive market for utility-scale solar projects worth GW. That is why the solar developer's head was reeling on that late July morning in 2019.

★★★

A lot of action had to take place during 2018 to enable the major breakthrough that came about in 2019. I was in the privileged position of being able to hold extended fieldwork stays in Portugal during this crucial period for solar energy development, spending a month there largely in August 2018 with a week in October 2018, followed by a month during February–March 2019 and another three weeks in July 2019. This allowed me to conduct numerous valuable interviews with a plethora of sectoral stakeholders: from grid distribution experts to energy journalists, from regulators to municipal energy advisors, from energy community practitioners to a range of solar developers, from energy researchers to energy association representatives, and from foreign investors to environmental activists. Additionally, I participated in the first-ever 'energy forum' during 16–17 July 2019, organized by a media agency, which convened many key players in recognition of the mounting interest in solar energy development in Portugal.

As before, I also made numerous field visits beyond Lisbon. Notably, one of these was to Evora, the seat of interesting developments on smart electricity grid infrastructure as well as concentrating solar power. Another was to Monchique on 26 July 2019, which had suffered Portugal and Europe's largest wildfire back in August 2018, one that took over a thousand firefighters a week to put out, while destroying 27,000 hectares of land, including 74 houses; of these, 30 were primary residences, given Monchique's attraction for holidaymakers to its thermal springs and wilderness recreation offerings.[1] This visit led me to write a popular piece for the *ECO123* magazine based in Monchique, which was translated from English to Portuguese and German, and received wide readership (since 2019, I have often been emailed by non-academic readers who came across this piece).[2]

Unlike my first fieldwork stay during September–October 2017, I came on these visits with a fuller understanding of Portugal and appreciation for the dynamics of its solar energy sector. In July 2018, a leading energy journal published an article I led based on my fieldwork from 2017 (Sareen and Haarstad, 2018) – this struck a chord with many researchers and was cited over 150 times over the next five years. This was useful to include when reaching out to prospective interviewees during fieldwork, and I found several of the leading actors in the Portuguese energy sector willing to make time to meet me, and to take the time to share their valuable insights and perspectives in a helpful and interested way. I tend to think this was not merely a function of my having become more embedded in the societal context, but also of growing recognition among sectoral stakeholders that the solar energy transition was getting underway in a significant manner.

In fact, by the end of my visit in July 2019, the buzz around Portuguese solar energy was inescapable. It was election season, with national elections to be held on 6 October 2019. The terrible wildfires of 2017 and 2018 had been compounded with the tropical storm Leslie in October 2018, being downgraded from a hurricane shortly before hitting Portugal's Atlantic coast, where 300,000 homes lost power, with wind speeds breaching 100 kilometres an hour.[3] While research shows that flooding is a relatively frequent occurrence for Portugal (Tavares et al, 2021), the spate of disasters exacerbated by changing climate patterns meant that 2019 witnessed the first real climate election in Portuguese history. An energy researcher based in Lisbon, discussing it with me years later, still vividly remembered the prominent climate messaging, remarking that it cut across party lines, with not just the winning coalition but all parties emphasizing that they would act on climate change. Another remarked that there was already a pronounced shift in the run-up to the European Parliament elections in May 2019, which had an impact on national political discourse.

This was not obvious until as recently as 2018. In fact, until October 2018, Galp Energia was trying to partner with the Italian ENI and pull together a consortium to undertake Portugal's first deep-water offshore oil exploration. I had a peculiar sense of entanglement across national politics when a friend in Bergen – where I lived at the time – mentioned that his colleagues at Equinor (the Norwegian oil and gas major rebranded from Statoil in May 2018) had been making trips to Portugal to offer technical support towards such a prospect.

It was one thing for Norway, deeply invested in oil and gas assets with a highly developed offshore industry, to be debating future offshore prospecting (a debate still underway in 2023). It was quite another for a country without any experience in this sector to embark upon a fossil fuel adventure starting out as late as 2018. The cognitive dissonance of such ambitions made itself felt in a country already facing the rising impact of climate change. The Campanha Linha Vermelha (Red Line campaign) from 2017 onwards was the most prominent series of organized protests against offshore oil drilling and fracking in Portugal.[4] Collectively knitting a multi-kilometre red line that should not be crossed, carrying out spectacular protests along the coastline, and gathering under the motto 'knitting to raise awareness and mobilize', this symbolic form of social resistance was instrumental in the consortium abandoning its plans and the government shelving its flirtation with becoming the latest country to join fossil fuel addiction in the age of climate change.[5] A year down the line, with the Socialist Party triumphantly entering a new term upon re-election, this courtship seemed a distant memory, as Portugal embraced the PNEC 2030 and RNC2050.

★★★

To explain the shifts taking place in 2018, I zoom to 14 August 2018, when I undertook a field visit to Estremoz. This small municipality hosted what had become Portugal's first subsidy-free solar plant, named Interventus 4 and located 5 kilometres outside the town. Eschewing the use of a taxi, I remember a pleasant walk along a relatively quiet road, taking in the lay of the land. I also recall a somewhat comic effort to hitch a ride back to town, drawing reprimanding looks and even a disapproving wag of a finger from an elderly driver who slowed down just to tell me off, before speeding away into the distance. Hitchhiking was apparently not an established way of moving about in the Portuguese countryside!

My visit to the town hall was enlightening in a different way. It was made clear to me that the municipality had no major role to play in the development, with no dedicated energy advisor. The officer who was assigned to me was the most knowledgeable available on the subject, and began by explaining that:

> 'The solar park has made no difference for the public. The only benefit is a 1.5 per cent share of initial investment, which is positive for the municipality. And maybe we can have the use of local clean energy in the future, which could be cheaper for inhabitants. We have two completed projects and two upcoming ones, and there are two long-term local jobs linked with each project. There were no local people employed in the construction phase, they did it with their own team.'[6]

While project promoters varied, one company was behind three of them. They ranged between 1 and 4 MW each, and had all procured land from the same rich landowning family, with some of the land demarcated as agricultural reserve being more time-intensive to repurpose. Proximity to the São Lorenzo sub-station and a chance to connect directly to a high-voltage transmission line, along with a simple land acquisition process, had made the process relatively smooth and efficient, from land acquisition in 2016–2017 to plants commissioned in 2018. Here, then, was a simple template for how medium-sized plants could proliferate across Portugal, with a few MW each being built close to the grid and energy demand centres, while expanding the transmission grid to make spatially efficient use of this modular low-carbon energy source that lent itself to spatial distribution. This would reproduce existing forms of social inequality, for instance benefiting large landowners with lease incomes, but would generate some local development funds indexed to solar project investments. If the share of project revenue were to accrue to the host municipalities and were to be meaningfully invested for local social impact, plants of a few MW could generate a tidy additional source of municipal development funds for small towns of a few thousand residents.

At the time, discussions at the European level were starting to gather steam when it came to renewable energy communities. This legislative and policy development would foresee a qualitatively different role for citizens in local renewable energy development, placing them in more active roles with regard to ownership and control over these energy resources and infrastructures. By December 2018, this process resulted in the recast of the EU Renewable Energy Directive (RED II), which led to Renewable Energy Communities being defined in the 2019 European Clean Energy Package (Lowitzsch et al, 2020). Emergent definitions, which needed to find national resonance in Portugal's PNEC 2030, explicated (co-)ownership by consumers in renewable energy clusters and communities, starting to play the roles of users and prosumers.

Interestingly but not altogether surprisingly, the small municipality had no plans – nor a particular strategy – when it came to the distribution grid concessions that would determine the ownership and control of critical local energy infrastructure. What made this especially consequential at the time, however, is that the distribution grid concession licences were set to expire nationally in 2019, having been held by the incumbent Energias de Portugal (EDP) since its privatization in the early 2000s, as Portugal transitioned away from a national public energy utility. Municipalities held the legal right to allocate these licences, and a process was underway under the aegis of the Energy Services Regulatory Authority (ERSE) to define these distribution grid concessions for the coming decades. All but the largest and wealthiest of the Portuguese municipalities, such as the big cities of Lisbon and Porto, lacked dedicated energy advisors and in-house competence on the technicalities of the energy sector. A few small municipalities had in fact run their own distribution grids for years, but the people in charge were ageing. Moreover, most sectoral stakeholders I met were of the opinion that the distribution grid lends itself to being run as a natural monopoly at the regional spatial scale, to harness economies of scale in data and energy infrastructures.

With a move towards digitalization of the electricity sector to enable two-way electricity flows and monitoring, for instance through smart electric meters (Geels et al, 2021), and the growing penetration of renewable energy sources, I expected that the distribution grid would gain enormous value. But I also recognized that it was unlikely I would find answers to the quandary of how this should take place here in Estremoz. Even if the first subsidy-free solar plant had come up in this small municipality in the eastern part of the Alentejo region, some key energy sector decisions were still run from Lisbon.

The conundrum of valuating a digitalizing electricity grid

I was delighted to be able to interview one of the most important former regulators in Portuguese history, now no longer working at ERSE. One of Portugal's most experienced energy researchers had recommended him to me during an interview on 13 August 2018:

'[He] critiqued ERSE for its lack of independent functioning, in parliamentary hearings. The regulations are endlessly complicated. I have quite some expertise and could not make sense of them after a whole month of study! There is complete opacity, ERSE is prone to play games with co-efficients, which can be manipulated. One priority should be to clean the mess. There is no real energy policy. Even the incumbent EDP does not like the Secretary of State for Energy, due to the hike in the tax component of electricity bills under Socialist Party rule due to the Left Bloc influence. The current state of affairs is a fight between the energy sector and the national government.'

So when I interviewed the former regulator, I was prepared for fireworks, and expecting a person who was not going to pull any punches. He did not disappoint.

The very next day, when we had a chance to speak, he narrated the arc of energy politics from the 2000s onwards. First came the very expensive bet on highly subsidized Amareleja. Then came some successes with wind energy, which led politicians to deprioritize solar projects and go for wind in a big way. Then the economic recession hit, the wind energy contracts backfired, and the International Monetary Fund and the European Investment Bank came calling. Their priority was to bring down electricity costs and to privatize the electricity sector, including E-Redes (formerly part of EDP) as the distribution system operator (DSO) and Redes Energéticas Nacionais as the transmission system operator. The government also had to stop subsidizing renewable energy, so while lip service to renewable energy continued, projects became caught up in bureaucratic delay. A cap on electricity tariffs in 2007, set too low, led to an accumulated tariff deficit of €5 billion by 2011. He sighed, then came in for the punch:

'The government had to manage growing tariff debt, keep prices down, so they dismissed subsidies on renewable energy. Then the current central-left government came back to power and maintained this line. This makes particular renewable energy forms very hard without any subsidies, especially as the Iberian energy market is distorted by Spain which constitutes 80 per cent. For instance, Spain forced generators to pay part of the cost of infrastructure as tax – conventional ones internalized this cost into their operations. The spot market works dysfunctionally, especially given our very high wind energy penetration. Now the government tells the solar industry they have to rely on this market price, otherwise they are not welcome here. So no rational conditions to attract any investor. Now we see some PPAs [power purchase agreements] being signed between some promoters and suppliers. They are strange investors, not the usual companies that invest in renewables. Why suppliers are engaging in this game is rational. Several feed-in tariffs for solar plants are expiring and they have to go to the market anyway. They have to learn how to trade and behave, so they are ready to pay a price to learn how to act under new circumstances. For them signing a ten-year PPA at €40 per MWh makes sense.'

He would have approved of the solar auctions as introducing a market mechanism to advance the solar market in 2019, and was perhaps surprised by the tenacity shown by the government when it came to developing utility-scale solar. Portugal showed willingness to move forward within the political winds of the late 2010s, and to carve out a path to usher in rapid solar energy deployment in the 2020s. But part of his criticism from August 2018 would stick for years to come:

'My criticism is that decentralized solar should be promoted instead of being stopped by the government, it is particularly strange for a central-left government. It should be supportive for local communities to manage their energy resources. Portugal is one of the few countries in Europe where you cannot do anything as an energy community even if you want. Even if you build your own micro-grid infrastructure, each household has to pay the full cost of energy. The politicians and ERSE are afraid that allowing this would lead to an increase in the distribution tariff, and anything that leads to even a small tariff increase makes the government panic. They have instructed ERSE to act in the same way. This is very sad because the residual issues can be addressed in simple ways. The government is blocking all kinds of innovation.'

Figure 4.2: Portuguese Parliament

He had read my early work, and gave me some advice and a challenge, adding that "your work can go further than the descriptive and analytical, and really make a difference by pushing for a public discussion to enable social innovation in the energy sector in Portugal".

When I read through the text of his intervention in Parliament, I was struck by his pragmatic push in line with this logic of enabling renewable energy communities, for spatially distributed forms of locally owned solar development. Figure 4.2 shows a view of the imposing Portuguese Parliament building, the sun poised behind it. He highlighted the valuation of the electricity distribution grid as a key challenge for Portugal, pointing out that the current concession was due to run out in 2019. His regulatory experience and understanding gave him the foresight to see that the grid would be an incredibly important asset for the energy transition that Portugal and the Portuguese people needed, one that would rely on people being able to invest in small local solar installations close to centres of energy demand. The digitalizing electric grid is a key asset for an electricity system increasing renewable energy in its mix (Trahan and Hess, 2021), as storage services become critical for energy flexibility with more solar energy penetration on the grid.

Sadly, and validating his critique, the distribution grid licence concessions were put off time and time again, and in the meantime EDP accelerated the rollout of millions of smart electric meters. The grid was digitalized, but it was the incumbent who decided the form this took, and by 2023, much of the solar content in Portugal came from utility-scale plants. As later chapters delve into, solar energy communities faced precisely the sort of hurdles the former regulator had characterized for the rollout of renewables more generally: red tape, delays in obtaining licences, being stuck in the space between being legal and being actually possible to implement with the blessings of the state. I took his advice from this interview to heart, and have indeed pushed for a larger conversation on enabling social innovation in the energy sector, concretely through large European projects on energy communities and on upscaling energy flexibility solutions based on local-scale innovations. But if one were to identify the biggest missed opportunity of 2018–2019 in the Portuguese solar energy transition, it is the failure to learn from Germany, the Netherlands and Denmark – all frontrunners whose experiences show the power of valuating the digitalizing grid as a key element of increasing distributed renewable energy capacity in a rapid and inclusive way. Portugal did not have the money to pump into the sector that these richer countries did. But it did have the opportunity to use the grid licensing concessions to better effect in 2019, if only it had fully grasped what it was worth to the incumbent – and to the future energy system – and used the courage of its convictions to act on this knowledge.

★★★

If one stakeholder took ownership of the narrative of the smart grid, it was EDP. I first visited their showcase demonstration project in Evora, EDP Inovgrid, on 27 September 2017. Along with a highly instructive and professional tour, I was informed I now had something in common with the United Kingdom's (then Prince, now King) Charles, who had been given such a tour in 2011. Inovgrid was a frontrunner project, piloting a smart grid for 31,000 households as early as 2010, which by the time of my visit in 2017 had expanded to a million households of the total six million in Portugal. This coverage had reached four million by 2021, with full coverage across Portuguese households expected to be completed in 2024.

The visit was enlightening in many ways for an energy social scientist. The sophistication of the digital energy infrastructure was a world apart from the many improvised arrangements I had seen in various countries, and the meticulous explanations revealed numerous points of interest when it came to how smart grids eased operations and increased efficiency at lower costs for DSOs. I made another visit on 13 August 2018, this time to see the results of some of their large set of European research and development projects. EDP had been perceptive about learnings from Inovgrid, noting that household interest dropped when they noticed that savings were quite small in absolute amounts, just a few percentage points. While this was significant for a DSO, it meant little to someone paying a few hundred Euro a year. Hence the ongoing projects focused on automation and aggregation, with the rationale that many households with solar production and battery storage could contribute valuable energy flexibility to the distribution grid. This is a money-maker for DSOs, as flexibility is an ancillary service provided at premium costs, where one could pool and trade many users' resources on the electricity market (Biegel et al, 2014).

Among the many fascinating demonstrations I was shown was a 360 kilowatt-hour (kWh) lithium ion battery, using stacks of 1.5 kWh batteries, in partnership with Siemens. My interlocutors also showed me a network of small solar panels and batteries hooked up across households in a village, with larger battery storage at the end of the lines, which they had been using to run experiments in island mode on this low-voltage grid. Based on these experiments and analysis, one of them explained:

'The only way prosumers could make money would be if they got a revenue share from providing grid flexibility based on storage, that is, if we could set up an automated algorithm for the network demanding power from the batteries at times when we are not producing enough wholesale power to match current demand. Then prosumers could get a share as the savings would be significant for the deviations avoided. This is also true with dynamic tariffs, but it has to be automated, users are not interested or aware enough to do it manually. With electric vehicles it is an opportunity too, as they will have an effect on grid use in rural areas in particular, by people with weekend houses in the countryside.'

For me, living in Norway, this was like having a future described as a parallel reality that I was already part of. By 2019, the majority of cars sold in Norway were electric, and I had been part of running a smart electric meter living lab with 46 households in Bergen, even as we reached 97 per cent of Norway's three million households on a nationwide smart grid (Rommetveit et al, 2021). Dynamic tariffs for households were commonplace, and I agreed with most of what I was told and what I saw thanks to these generous Portuguese engineers. Yet I also had a clear sense that all of this could either be done in a manner that was citizen-centric, or in a way that maximized profits for the DSO or other intermediate actors.

Thus, at the end of 2019, even as the Ministry of Environment and Energy Transition declared that cost-competitive solar energy would play a crucial role in alleviating energy poverty, and while the mixed feelings and charged emotions of various stakeholders came to the fore, I felt the sense of an opportunity lost, or at least delayed. As the PNEC 2030 explicitly mentioned a target of energy poverty alleviation, and Portugal began to enact its ambitious vision of decarbonizing electricity and electrifying many sectors in a bid to achieve a low-carbon vision among the most ambitious in Europe, I dwelt upon the great deal at stake. The stage was set for ambitious cross-sectoral policies to enter messy implementation. Yet the outcome, I felt, would not necessarily benefit ordinary citizens.

FIVE

Solar Portugal 2020–2021

The best-laid plans and the mixed news of 2020–2021 for Portuguese solar

At the start of 2020, as Lisbon took over as the new European Green Capital, few would have imagined how the next couple of years would look. As luck would have it, from March 2020 onwards, recognition dawned in a widespread manner across Europe that the COVID-19 pandemic was a worldwide calamity that would take years to address, and to recover from in the fullness of its adverse societal impact. A year down the line, when I coordinated a hybrid training school based on a five-country hub model in April 2021, the Portuguese trainers and trainees gathered in Lisbon joined online wearing face masks in front of their individual computer monitors. The pandemic circumstances had become the new normal for a protracted period, and life went on. Thankfully, by the end of 2021, rapid vaccine development and easier (albeit globally highly inequitable) availability had brought things back to a more familiar way of functioning. Even so, there were parallel emergencies, and the training school we convened thanks to the ENGAGER European Energy Poverty network addressed the timely topic of mainstreaming innovative energy poverty metrics.[1]

Since the prominent mention in the National Energy and Climate Plan 2030 of 2019, energy poverty had gained major traction in Portuguese public debate, and crucially, also a foothold in energy policy (Horta et al, 2019; Mafalda Matos et al, 2022). The country had impressive research capacity on important aspects of household energy poverty, which allowed the development of a database and typology of energy efficiency and thermal characteristics of buildings, downscaled to the parish level for the country (Palma et al, 2019). This was quite an accomplishment, given that Portugal has 3,091 parishes, a lower spatial and administrative scale than its 308 municipalities. Researchers also took into use new data streams, such as smart electric meter data (Gouveia et al, 2018), to identify and address the multi-dimensional problem of energy poverty. So hosting a hub of the training school in Lisbon was appropriate, given the innovation on metrics of energy poverty taking place in this part of southwestern Europe.

Energy poverty is most influentially defined as 'a condition wherein a household is unable to access energy services at the home up to a socially- and materially-necessitated level' (Bouzarovski et al, 2012, p 76). It was initially recognized and defined in the United Kingdom and Ireland, but only from the mid-2010s onwards did other European countries start to embrace it as an object of research and policy. Discussions beyond Europe tended to focus on basic energy access in more subsistence contexts, and certainly there was also the need for a parallel conversation about energy over-use by a consumerist class of elites. But what was heartening – and timely – was the emergence of a focus on energy poverty in European countries like Portugal and Spain (but equally, Poland and The Netherlands, and more), facilitated by the ENGAGER network. This brought together thematic researchers and practitioners, thanks to funding from the Cooperation on Science and Technology Association during 2017–2022 (Jiglau et al, 2023). Given the contrasting geography and climate of Portugal, innovation on energy poverty metrics in this setting also meant greater attention to aspects such as energy for space cooling, with climate change making summers in southern Europe intolerably warm and posing health risks in built environments maladapted to such high temperatures over increasing seasonal spans. During training school discussions, there was knowledge exchange on the scope to impact practice and bring new metrics into national protocols, creating a knowledge base to actually tackle energy poverty. Some of the participants were instrumental in establishing and leading the European Commission's Energy Poverty Advisory Hub, an initiative to alleviate energy poverty in sync with accelerating a just energy transition at the municipal level.[2] As expressed in an article led by one of the trainees (Mahoney et al, 2022), energy poverty had to be addressed in conjunction with energy transition and climate action measures, in order to ensure an integrated action agenda that tackled these issues in a coordinated way for a just transition.

The COVID-19 pandemic presented considerable challenges, including exacerbating energy poverty as people in Portugal were constricted to their households most of the time during 2020–2021, but it also offered leverage for policy action to step up efforts to address energy poverty. International networks, fostered through collaborative infrastructures, helped create the conditions to inform policy change, for more streamlined energy poverty measures to emerge. Not least, activities like the training school exemplify the work underway to create the new competencies required to steer and implement just energy transitions. Research has a key role to play in this respect. Despite the considerable challenges and gaps that the pandemic circumstances created for energy research (Fell et al, 2020), in this instance progress was underway to develop and implement energy poverty alleviation measures. This was much needed, given the high incidence of household energy poverty in Portugal.

★★★

Figure 5.1: Large solar plant construction in a rural landscape

As pandemic measures stalled numerous aspects of everyday life in Portugal, there was concern that the second solar auctions, announced for spring 2020, would be indefinitely postponed. The government did move them a bit down the line to adjust protocols, but to its credit, managed to conduct reverse e-auctions for 670 megawatts (MW) in August 2020. Figure 5.1 shows a view from above of a large solar plant being built in a rural landscape. Not only did it implement these second auctions, Portugal set yet another world record, this time for the jaw-droppingly low tariff of €11.14 per megawatt-hour (MWh). This was criticized for being too low to be feasible, yet several other bids also featured very competitive tariffs. Solar was here to stay.

Be that as it may, the COVID-19 pandemic did disrupt several aspects of the solar value chain. This is a highly globalized chain of commodities, that come together from predominantly China but via various locations during extraction, assembly, shipping, and so on (Mulvaney, 2019). As many parts of the world entered indefinite lockdown, the solar developers awarded contracts in the 2019 auctions – as well as those who won auction lots in 2020 – faced unforeseen supply crunches that challenged their ability to get plants commissioned in time. The government was understanding of this, and allowed standard annual extensions beyond the original deadline of two years to install photovoltaic (PV) capacity from when it had been auctioned. Some projects did indeed fade away in the interim, but most stayed on track, albeit only to come up along adjusted, more relaxed timelines. When I interviewed industry stalwarts in February–March 2022, they were confident that most of the auctioned capacity would come on grid shortly, and indeed projects did. In addition, massive solar projects based on power purchase agreements also progressed, notably the mammoth 220 MW Solara4 Alcoutim solar plant.

When the solar plant in Alcoutim, developed by a Chinese-Irish consortium, was commissioned on 9 October 2021, it became the largest unsubsidized, privately funded solar development in Europe. Some years behind the initial schedule, to be sure, as some had anticipated completion in 2018, but a major landmark nonetheless. I recalled a solar developer gazing out at the landscape along with me in 2017, and remarking that the undulating landscape was ill-suited to solar development and that the project would prove too costly. The developers had certainly proven such naysayers wrong. With a tinge of sadness, I also recalled an interview with a representative of the civil society organization Almargem Association on 20 September 2017. This organization had expended considerable effort and resources on documenting the 300-kilometre cultural and natural trail 'Via Algarviana' that cuts across the breadth of the Algarve from Alcoutim to Cabo de São Vicente.[3]

A representative was dismayed by the prospect of this large solar plant cutting through this documented heritage and asset for regional tourism. He elaborated:

'The EIA [environmental impact assessment] responded to our comments protesting the solar park being located in Alcoutim, suggesting adding solar panels as one of the highlights along Via Algarviana, or making a detour in the Via Algarviana. The developers never spoke with Almargem Association, we spoke with the Alcoutim municipality. The developer laid the foundation stone in early April 2017, with the Minister of Economy from Lisbon in attendance and other bigwigs from Faro. That was the first time the Chinese investor saw the landscape, I think, then tried finding alternatives, but the municipality didn't talk any more with us. It was APA [the Portuguese Environment Agency] in Lisbon that was involved. In Portugal, the APA responds to a developer proposal by working many publicly paid hours to suggest an alternative proposal that the developer can then easily adopt, and it's already accepted before details become public.'

Through the municipality, the organization had sought a public discussion with the company, which had not shown up. The representative held the position that small solar plants near villages with significant populations for nearby energy demand were desirable, and that people should be encouraged to install solar solutions in their own households. But this, my interlocutor recounted with an air of curiously combined frustration and resignation in the scenic town of Loulé, was not how local decision-makers operated. He continued:

'Municipalities do token gestures for panels in villas, but no work to promote small distributed solar. They get project proposals and first ask what they can get out of it, trying to create local jobs even when this does not make sense. Large projects constitute an easy investment, less work for the municipality, so they bend over backwards. There is usually no environmental policy, it is just namesake.'

And yet, things were more complex than this, as he went on to explain:

'20 years ago the government used to do a shoddy job so it was easy to critique, but now the municipalities and the national government agencies have become professionalized and hard to counter due to their expertise with many well-paid specialists. It is conventional politics with no regard for environmental issues when a project comes.'

His critique was that while the form had become sophisticated, EIAs amounted to little of substance, as the process was set up as a way to ensure project progression. The garb was a mere device for social legitimation, a way of 'sustaining the unsustainable' (Blühdorn, 2007, p 251), of rendering the matter technical (Li, 2007), beyond the reach of local activists.

On 26 July 2019, meeting another socioecological activist in Monchique, I was reminded of the Via Algarviana. He talked about having guided many tourist groups along the ancient pilgrimage route with an air of reverence. They would retrace the route the pilgrims followed from Alcoutim at the Spanish border towards the promontory of Sagres in the southwestern corner of Portugal, where they found the relics of Saint Vincent (Vincenzo). These were brought to Lisbon in 1173, and in 2023 the city celebrated the 850 year jubilee of its patron saint with great pomp and splendour in a society reopened for festivity after the pandemic years. He emphasized the importance of slow tourism, encouraging visitors to spend time to take in the landscape in a deeper sense. Very concerned about personal carbon emissions, which he had been tracking closely through a tool he was popularizing for others as well, he remarked ironically that:

'We would have to electrify the whole stretch of land in the south of Portugal and strengthen 200 kilometres of railway lines to manage our 80 per cent tourism-dependent economy here in the Algarve [in a low-carbon way]. [Otherwise,] if we decarbonize by 2050 and flights are very expensive, then how will tourists come?'

★★★

The constraints imposed by the societal emergency of tackling the COVID-19 pandemic meant that many sectors – like tourism in the Algarve region – took a beating (Santos and Moreira, 2021), and needed novel forms of support using national and European funds. Extraordinary times called for extraordinary measures, and Portugal's Recovery and Resilience Plan kicked in as an essential support measure. Through this mechanism, the European Union allocated 13.9 billion Euro in grants and 2.7 billion Euro in loans to the national economy during 2021–2026. Through billions of Euro in subsidies to fossil fuel suppliers, Portugal did manage to keep energy poverty at bay. What is more, it even managed to exit coal, a full nine years ahead of its 2030 schedule compared to its commitment as part of the Powering Past Coal Alliance, with the shutdown of Energias de Portugal (EDP)'s last coal thermal plant at Sines. This freed up a massive 1.2 gigawatts (GW) in transmission capacity on the high-voltage grid south of the capital, creating a vacuum that utility-scale solar would flock to fill after 2021.

As the Portuguese population switched to work-from-home mode, one of the people who had been closely tracking the trend of utility- and lower-scale solar in Portugal started to increase his engagement with online platforms and software applications, in order to pool the many haphazard online data sources about this phenomenon. This led him to found the Observatorio Fotovoltaico (PV Observatory).[4] When I interviewed him on 11 July 2023, he recounted a fascinating story that sums up many of the defining characteristics of the Portuguese energy transition. He had an early interest in solar energy in Portugal, and years ago, was frustrated to find that only annual figures on national PV capacity were publicly accessible via the Directorate General of Energy and Geology

(DGEG) website, with very little disaggregation (subsequently, statistics became available on a monthly basis). He put it more bluntly:

'A PDF [file] with a table inside with poor formatting. So either there was very large-scale and abstract information, or very detailed for specific cases but with large gaps. Then a colleague in my lab had an automatic software to detect all installed solar PV based on satellite imagery.'

He soon cottoned on to the fact that detailed statistics were available, but only for sale to businesses. He was quite active on the professional networking website LinkedIn, and noticed an abundance of details about specific solar projects, which he began to catalogue manually, including links to sources with the contextual information. He reminisced about this:

'It was painstaking and slow work, manually using Google Earth, covering nearly 500 installations. I felt like a stamp collector, doing it without a clear motivation, just hooked. Then after three years, I started to use two different mapping applications. But they were not evolved enough. Then I came to France, and a colleague introduced me to an open access French software, more interactive and functional, with filters. If I chose a filter and pressed three buttons I could obtain an interface that would be easy for anyone to use.'

He stated with pride that Coopérnico, Portugal's first solar energy cooperative, used his map on their website. He voiced a desire to make a publicly accessible, user-friendly map of solar installations, and the need to balance this with financing his time spent in creating it. So far, this had been driven by goodwill and interest, but companies were showing interest in having versions customized to their visual profiles. This could present synergistic prospects.

He recounted how, using his network, he went through Lisboa Enova – Lisbon's municipal energy agency with its own active solar promotion programme called Solis – and received help in coding, to optimize the extraction of all online information on specific projects. He explained this scraping technique succinctly:

'Companies have a code that describes their sphere of activity, and I can run a code that can find this code for all companies. I have covered over 1.6 GW, so more than half of all 2.6 GW currently installed in Portugal. And now I can see patterns for this new sector, for instance that a lot of the PV that has come online is linked to the food industry.'

He was preparing to promote an agrivoltaics prototype – a PV plant co-located with agriculture – at the University of Lisbon the following week. "The goal is to share information with everyone as the visual interface, but I need to study some details before sharing the raw data. Then I want to link it to other demographic details." He could not include residential solar within this purview, because of limited public information about it online, and because European privacy regulation imposed limits. But importantly, also because this made for poor conversion of time invested to increased coverage for his tool in terms of kilowatts. By contrast, commercial installations were always prominently advertised online, with the clients and details, and usually came in larger chunks per project.

He was bubbling over with ideas, and I was struck by how a freelance enthusiast could come up with so much, whereas the well-resourced executive agency, DGEG, continued to run a quite staid website with limited functionality for ordinary users. What a missed opportunity!

And he wasn't done. He continued:

'I will do analysis for Portugal on how watt-peak per module evolved in Portugal, whereas now the watt-peak is very high and similar across solar projects in Portugal. So I am very open to various types of collaboration. Maybe industry matures and wants to buy data, maybe researchers will do joint analysis. Maybe I can get access to data like tilt module, inverter capacity, to understand the relationship between peak capacity and inverter capacity. This is too private to make public per project, but aggregate analysis can be of interest.'

He reasoned that by providing concrete examples, the observatory could inspire people: "For instance, if I can show that a lot of municipal swimming pools have solar PV and that helps others adopt it. Or floating PV projects. I am now starting conversations with energy agencies across the country to help them to identify potential projects." He envisaged all municipalities embedding his map using a simple location filter, to track installations within their territory. Resonating with others who had recounted similar experiences, he knew that municipalities – their political power notwithstanding – also struggled to get the data they needed from DGEG as much as any other actor.

Clearly, the lack of initiative and responsiveness from DGEG had provoked him to be persistent:

'A motivation for my project for the past four years [since 2019] is that these detailed tables of information on solar installations exist at DGEG and this is public information. I called them every two weeks for eight months to ask for this data, and they always said that it will come. One month after I contacted one person at DGEG whose email I found, I got an amazing dataset, characteristics for all installations until 2020. I used [the software] Power BI to analyse the data, you could address specific questions using the data, regarding the spatial distribution per district. Then I asked DGEG if I could get regular data, but they said only once annually. So I followed up after a year and they gave me one more update, but not at the municipal level. They have a lack of interest and time. They do not see these small personal initiatives as meaningful. They asked me to remove their DGEG logo where I credited them for having provided the basic data. This is the motivation for me to scrape online data. Agencies like DGEG and even APREN [Portuguese Renewable Energy Association] say they support such efforts, but they do not open up basic data systematically, even though for DGEG this is public data, just not publicly accessible. Last week a company contacted me to validate if their rooftop project (6 MW) is the largest nationally. My LinkedIn post later this week will be about the largest self-consumption plant, 17 MW, supplying a single paper factory.'

With his analytical sensibilities and his long-term interest in solar energy sectoral trends, he could see that fine-grained data, handled well, could turn individual solar projects into more than simply points of data: "they can capture the imagination with particular details". He hoped to do a book about the history of Portuguese PV. He wished me luck with mine, and I can only hope that his will also see the light of day, and shed further light on this complex story. He had been archiving Portuguese media commentary on solar energy for years, and imagined the power of linking each solar project to its EIA, to help people recognize patterns within the sector and the narratives put into play, and to cross-reference this with other land uses to extend his database.

None of this existed during 2020–2021; it was coming together in his mind and through his work with Observatorio Fotovoltaico in 2023, as it started to take off. I highlight this valuable initiative and perspective because it shows the power of simple good ideas, with some smart application. As solar energy researchers like him and me are painfully aware, such solutions do not exist – and are not championed top-down by national agencies – in this burgeoning sector. They have to be pieced together by proactive individuals, and to some extent the work of industry organizations such as APREN helps pave the way for this. But the main sectoral authorities continue to regard and run the energy sector as a closed, centrally managed system, not seeing the value of empowering ordinary citizens and enabling their closer engagement with solar development towards just energy transitions.

This sort of work holds a glimmer of hope that energy democracy can in fact add tools to its kit for an empowering form of solar energy development that puts people at the centre, even if the push comes not from a benevolent government or its seasoned bureaucrats, but from a younger generation that does not share obsolete assumptions about a sector whose fundamentals have changed, and through individual agency born – simply – of impatience.

Finding a balance between slow and fast emergencies

There is a sharp contrast between the handling of the COVID-19 crisis – a *fast emergency* – and the *slow emergency* of climate change, of energy poverty, of socioecological injustice (also see Anderson et al, 2020). When push came to shove, Portugal displayed remarkable resilience, an ability to soak up pressure and to rally its limited resources as a society to stand up for and protect its own. It had very limited hospital beds, yet it hunkered down and took the precautions necessary to ensure that vulnerable people – the elderly, the ones with comorbidity concerns like asthma – were able to access the support they required, in their hour of greatest need. This is not altogether surprising in a culture known to be risk-averse, which for instance has some 800,000 people accessing the socialized electricity tariff (Mafalda Matos et al, 2022), which allows a basic level of affordable electricity access to those with a limited household income. This is a remarkably high level of social provision premised on consideration of equity and a basic minimum.

Yet, this care economy does not quite extend across the spectrum of *slow emergencies*. Despite solar energy having reached grid parity in the mid-2010s, Portugal did not jump at this opportunity to bring all households on board with a solar energy transition that could reduce the debt burden on an ailing energy sector and accelerate national energy transition in an equitable manner. Instead, it hummed and hawed, took a conservative and circuitous route. Some might argue it is in keeping with a risk-averse cultural proclivity. But in my view, this is more a reflection of embodied incumbency in the key decision-makers of a conservative *sector*, one too used to being steered in a centralized and top-down manner. Similarly, when it comes to the electricity distribution grid concessions, Portugal continued to dither and postpone them indefinitely. As late as 2023, the government made agreements that would push the licensing concessions into 2024, extending the period with EDP running the grid. For perspective, EDP's licence for the lucrative Lisbon municipality was first set to expire in 2017, meaning it had been extended piecemeal throughout my entire tryst with Portugal!

These bureaucratic and technocratic obstinacies, this reluctance to change old ways of doing things, or even to assess alternatives from a more neutral vantage point than the comfortable confines of path dependency, routinely typify energy sector decision-making. Publics are starting to wake up to these deep-seated tendencies – energy social scientists have certainly done their share of shaking heads, and energy community activists have shaken their fists – but to little avail when it comes to changing the way things are done in corridors of power. As I attended the tenth jubilee celebrations of the European Community Power Coalition at the European Parliament in Brussels in September 2023, I thought about how far things had come, but also curiously, how similar things had stayed, despite all of it.

★★★

Signs of the increasing importance of solar energy began to be embedded in the landscape. Figure 5.2 shows an art installation by António Jorge Gonçalves along a Lisbon street, with a

Figure 5.2: Art along a Lisbon street that features photovoltaics in depicting the future

comic panel along the same strip showing PV as part of his exhibition 'Yesterday I saw the future'. In January 2022, EDP profiled an art exhibition by Caroline Piteira at Central Tejo (an old power plant in Lisbon), with the theme 'Out of time', where the artist focused on renewable energy sources and their connection with time.[5] That year, the national headquarter building of Ciência Viva, the National Agency for Scientific and Technological Culture on the eastern edge of Lisbon in the upmarket business district near the Oriente railway station, featured a striking installation of a 'solar tree'. Within the circular ensconce of the impressive museum-like building stood a large white tree-like structure, with solar panels radiating out as branches. An artistic rendition of a tree of light?

On 28 February 2022, I was able to meet with some of the most important staff of Ciência Viva. They were setting up a network of science 'farm' centres – 'Quintas Ciência Viva' – across Portugal, and mentioned a plan to establish a solar-themed one in Moura. I was fascinated by this idea. These farms were meant to be contextually specific as a form of science outreach in smaller towns around the country. Some were themed on regional produce such as olive oil, cherries, wine and salt. "Our aim is to have a small space for engagement and communication in the rural areas where there is a spirit of modernity and innovation", my interlocutor informed me. They partnered scientific institutions, and importantly, local municipalities, with 21 initiatives underway, involving finding funds and building facilities. The settings ranged from the grand pavilion of knowledge in Lisbon where we met, to an installation in a church in Tavira. Public schools would visit, they would host visiting scientists, drawing on 22 years of institutional networks, including over 400 local science clubs. The aim was knowledge dissemination, but also to promote social cohesion, bring actors together and serve a productive purpose locally.

As they spoke about the plans for the solar science farm in Moura, my head raced. The Ciência Viva Sun Farm would combine art and science, and include tactile elements. "We will emphasize sustainability, light and solar energy as a universal resource. And then the more technical ideas about what the PV technology is." They showed me a slide deck, with carefully compiled visuals such as how solar technologies could be displayed to visitors, extending them into the urban environment, and drawing attention to recycling panels. They had met with academic institutions and companies, the municipality, and held two meetings "more on political aspects, to get things off the ground".

At this point I had to ask: had they looked into the old PV manufacturing facility in Moura? What about the political economy of the developments linked to the Amareleja solar plant? Could they imagine telling this story along with profiling the technological aspects? Some of this was news to them, but they took notes to bring this up in a future meeting. "We should hire you as a consultant to make the plans!", one of them exclaimed.

'I went twice to Moura and met with local political leaders, but in the schools and general public, I do not think they are really aware what happened in the last years. Of course it benefited the municipality to have the big solar plant, but it remains quite aloof from local people. So we want to involve people, spread this consciousness about sustainability to the public.'

They wanted to make explicit links to solar vehicle charging, and to bring in examples of other solar plants around the world, to show varied applications. They were mindful that "this is an audience that also needs to understand how big the world is, some have never even seen the sea. We will have workshops and a solar lab that works like a makerspace".

Then came a remark that was fantastically perceptive and also summed up the difference between the capital and rural places like Moura, the contrast of pace and slowness, that permeates this chapter. My interlocutor shared a poignant reflection. She said:

'What happens in Amereleja is Alentejo speed, slow, much slower than Lisbon. Our role is this science communication approach. We work with limited capacity at the local scale. Amareleja is stopped in time many years ago, and they are there. We have ideas to install solar aesthetics and devices in urban settings. Nothing like that exists in Moura right now.'

There it was in a nutshell. Part of the challenge of solar energy transitions in Portugal was in creating societal understanding and social movements, in reaching into the rural regions and instilling a sense of ownership, of urgency, to plumb the depths of locally relevant options.

SIX

Solar Portugal 2022

A legislative basis for community energy in 2022 and slow solar growth

Things do move faster in Lisbon than in the hinterland of Portugal. During 2017–2023, the real estate market accelerated exponentially. Even so, most things do not move all that fast. I recall sitting writing at a blues café I was fond of, not far from the beautiful park of the Gulbenkian Foundation, when the grocery store owner across the road walked over to the house next door. Using rope, an elderly man lowered a basket with his shopping list and cash, and the shopkeeper filled the bucket and put in the change, ready for his regular customer to pull the rope and bucket up to his third-floor apartment. This was their little social interaction, an informal support network in a neighbourhood where people lived.

That was on 4 March 2019, when my writing companion was a researcher who was deeply concerned about the neoliberal forces taking over Lisbon, forcing various vulnerable people out of their residences in nice neighbourhoods of the capital because they could no longer afford to live there with rising costs. Many apartments had been taken over for short-term rentals catering to a rich tourist segment, further inflating the housing market.

I had this in mind when submitting a grant application with various European partners in 2020, which involved a social housing project in the neighbouring municipality of Almada, just south of the Tagus river. The New York State Renewable Energy Development Agency had devised a Sun4All model of promoting pro-poor low-cost community solar energy. They would soon begin to expand it to 800 megawatts (MW) across the state, giving low-income households cheap virtual solar electricity sourced from publicly installed solar plants on communal buildings such as public schools. This enabled savings on their monthly electricity bills compared to standard tariffs. In our Eurosolar4All project, we wanted to pilot a similar approach in diverse legislations in Europe, including for a social housing cooperative in Almada in Portugal.[1] When this received European funding via the Horizon 2020 programme, I was delighted to be part of undertaking the social impact assessment of this and other pilot projects during 2021–2024.

I first visited the social housing cooperative on 23 February 2022, accompanied by an energy expert and Eurosolar4All colleague from Almada municipality. An energy modeller joined us, who was combining multiple datasets to estimate viable solar installation options at high spatial resolution. He was a member of the solar energy cooperative Coopérnico:

'I'm trying to plug an information gap in terms of information about different locations where this is feasible close to energy end-use. And the techno-economic models at present do not create incentives for small actors to participate in the sector. So there is a need for pilots that show how smaller solar rollout can be done, and for national agencies to set better terms to make this happen.'

The choice of this cooperative was strategic in this respect. It was in the main social housing district of Almada, with some 5,000 households, about half owned at the municipal level and the rest at the national level. They had worked with energy efficiency retrofitting interventions here from 2013 onwards, choosing an apartment block where the residents had been relocated from Costa da Caparica 10 kilometres away, where their squatter settlement by the sea had been deemed vulnerable to climate change and taken down. They had a sense of social cohesion when they moved into that cooperative block built in the 1990s, for very low rents.

My interlocutors showed me the interventions implemented over the years, from facades with insulation to double-glazed windows and solar thermal water boilers. They had installed 4 kilowatts (kW) of solar panels, and by then these had generated 23 megawatt-hours (MWh) beyond heating the water. This surplus was injected directly into the distribution grid, without any compensation from Energias de Portugal (EDP), "as we did not manage to pursue the registration at that point and the remuneration would have been minimal under rules then", said the municipal worker. The heated water supply was a municipal service, for which meters had been set up for individual apartments: "This was the best way to do it due to bureaucratic constraints, but it meant installing second water meters for each apartment, costing €3–4 extra monthly." Residents had disliked this, but then saw savings in their gas bills, while they did not have to pay for the solar water heating.

The solar panels on the rooftop were barely visible, even after crossing the street. We did climb up for a closer look at the control unit. But replicating this required project funding:

'The idea with this project is that it can be replicated across other social housing blocks, or elements of it, through projects like Eurosolar4All. We get funding for such activities mainly through European projects or indirectly through structural funds handled at the national level. But a major problem is political change within the municipality, not the party in power [which was the same as in Lisbon] but the specific people. The municipality wants to close down the Almada energy agency as a separate entity, and to also dissolve two other thematic departments pertaining to building use and to entrepreneurial incubation. They want to do these things in-house, but no clear alternative plan exists after 2.5 years. So Almada has gone from being very successful and even a frontrunner on some environmental aspects such as mobility transitions and building energy efficiency, where we used to be at par with Lisbon municipality and well networked internationally, to having no buy-in at the political level and no clear horizon for energy project planning.'

The plans my interlocutor explained they had been promoting made good sense: to identify large public buildings with high electricity consumption and push rooftop solar investment for those, then scale out to neighbouring communities within the new Renewable Energy Communities (REC) legislation.

He also mentioned the municipal cultural centre, which I would visit the following year, the very week their rooftop solar panels began operating in June 2023. Figure 6.1 shows this rooftop solar plant installed in Almada. "But the social fabric there is high income", he said, reflecting that "we would like to use part of the benefits from that solar for other actors, such as these social housing units". That was not yet possible in 2022, but they wanted to explore options like virtual accounting if it had positive social impact. They had been frustrated by unwillingness to engage with communities among some actors they had approached. Then internal delays and lack of political will had doomed the traction they had generated with local stakeholders. They also struggled to get access to national authorities like the Energy Services Regulatory Authority (ERSE) and the Directorate General of Energy and Geology (DGEG) for guidance:

'They are hard to get hold of in practical terms, even though they are very open when it comes to participation in webinars and so on. The competence for managing solar plants should be at the local level, right now it is done nationally. ERSE is increasingly asking municipalities to take charge of this. They just don't have the capacity.'

The energy modeller summed up the deadlock. The government vision focused on big projects, not the medium scale in the urban environments. He wanted to see plants of a few hundred kW, a few MW, not necessarily on rooftops but on nearby land. He mused that a recent ministerial interview had taken the line that the government could allocate a certain volume of grid capacity, but could not decide the siting of solar projects. This mixture of state-led and market systems left companies to decide where to build, "who go for the cheapest option". The municipal colleague added his agreement: "So then you do not get the best outcomes in terms of fit-for-purpose. Why should we have a big solar plant on an area that should be used for agriculture?" The energy modeller added wryly that design principles were lacking. If cost-benefit analyses considered real needs and trade-offs, large-scale solar would not always have obvious economies of scale over a more optimized solar rollout at the community scale, with benefits of efficiency and advantages across sectors.

Figure 6.1: Rooftop solar plant at the Almada cultural centre

We sat outdoors by a café with a grand view over the beautiful municipality and nature. The conversation turned to the distribution grid concessions. At the start of 2022, the new policy had finally come out, with ERSE proposing a single national concession, which would do away with all possibilities of local ownership. They reflected that the public sector would lack the capacity to run the distribution grid, and it would cost a fortune to buy the infrastructure off E-Redes, the renamed distribution system operator, which had the know-how and the experienced workers. The municipal worker mused:

'There is money to be made, there would be a huge payoff. We have no problem running our water services in Almada. Historically in the 1950s we had public electricity services. Whether the investment is meaningful is one thing, whether we can afford it another, but there was no systematic evaluation. There was some talk about it but nobody did it. The national grid has huge benefits that enable optimized operation. So it closes all other options.'

It had become evident to them that RECs would only be a small part of the 9 gigawatts (GW) of solar capacity Portugal was targeting by 2027. Unlike a large company investing big capital, municipalities did not have an instruction manual or strategic business plan. They were concerned about poor people, because energy transitions that would benefit these had to deal with financial constraints, not just expect households to invest. They saw the necessity of municipal agencies building bridges between technical installers and public buildings to operationalize national energy policies for local public entities, helping them avoid some market instruments to prevent poor investments.

The energy modeller recounted a political debate:

'[T]alking about lithium extraction as being bad if done by companies for profit, but fine if the state did it. But if energy could be seen as something other than a commodity then we could go beyond such a valorization. We need different layers of governance to work on energy, where cooperatives and local energy agencies should be at the core of those who drive the energy transition.'

The municipal worker chimed in:

'[E]nergy has not been part of the core business of the municipality, whereas with grid concessions you get virtual money. Now we have a government that is decentralizing many sectors like health and education, but is pushing for more centralization of energy. So it never comes up in the discussion on decentralization.'

So what could be a game-changer? On this, like so much else, we agreed: net metering would make people very interested in solar plants at lower scales, and kickstart RECs.

★★★

On 26 June 2023, I returned to Almada, this time meeting with six residents of the social housing cooperative, along with other Eurosolar4All colleagues who translated between Portuguese and English. Five of the residents I interviewed were generally favourably inclined to the effort to make pro-poor solar possible. Four who were women spoke about their access to electricity and everyday use of energy in the buildings, being largely pleased with living in the social housing cooperative. The last two whom I interviewed were both middle-aged men, the first mild-mannered, the other irate. While one made his points gently and the other was emphatic to a fault, their chief concern was crystal clear. Despite two years of the project, and even longer lead time in one building where the rooftop solar I had seen in 2022 had been installed years ago, residents had seen no benefit on their bills from the local solar energy production.

They had been patient, and understood this was not the fault of the municipal workers, who were doing their best to implement the model. They had seen the energy efficiency retrofit interventions create benefits for those who resided in that cooperative building. Yet the whole idea of reducing electricity bills by partly consuming locally generated solar had remained just that, an idea, both in the building where a plant was in fact producing electricity and feeding it to the grid, and in other buildings that were still waiting for rooftop solar to be installed.

The irate interlocutor landed his critique, which was insightful, despite his agitation:

> 'Because not many people are coming to the sessions, maybe it is not worth installing solar panels here, because people are not interested enough. People will not see the value of the municipality offering them energy from solar panels, they do not care about the project as they do not believe it yet, they do not understand it so they are not interested, some just come out of curiosity. Maybe if the government makes it mandatory for solar panels to be installed on rooftops then it will work. But people are not interested in sessions about optimizing their energy use with washing clothes, they already know how to do such things. Eurosolar4All will fail in a place like this social housing neighbourhood. The investment in solar panels will not be locally valued. So there is a social challenge. Even if some buildings are compliant, it will be difficult to replicate in other similar neighbourhoods.'

His critique was partly aimed at the nature of the very places that needed public support. But it was also aimed at the modality of offering a service that took so long to implement. To some extent the municipal workers agreed with him; they shared his frustration at having been held up. Both the municipal bureaucracy on procurement and releasing funds took time, and so did approval from the national agency, DGEG. This only came in July 2023. Finally, two years after the injection of funds from a European project, one small solar plant had received approval to be installed to benefit residents of a social housing cooperative.

★★★

This was not a standalone story about the barriers that slowed down the rise of RECs in 2022 in Portugal, despite the regulations finally being in place – at least in principle – to implement them. There was a great deal of interest and several projects. I visited one called Smile Sintra,[2] near the picturesque town of Sintra whose scenery draws tourists in droves. My interlocutor on this occasion drove me there from Lisbon, chatting along the way to Bairro da Tabaqueira, a village with a peculiar history. In the early 20th century, the industrialist Alfredo da Silva – perhaps the richest Portuguese person at the time – had established large manufacturing operations in this village. Despite controversial links to the Estado Novo dictatorship, his larger-than-life statue still towers over a square; he had been known for treating workers well. Today, it is a sleepy place, and the evening of my visit, the residents who turned up for the project meeting at the local community centre were quite elderly. The setting we were in had been customized for working-class workers specialized in the tobacco industry back in the 1960s.

The meeting was for the Smile Sintra project to present a proposal for a 50 kW solar plant to be built on the public school, to start an REC with members from the village. The proposal was based on trained students having conducted energy audits of several social housing households. A community engagement expert from the Aga Khan Foundation had worked diligently to ensure an inclusive process, and I noticed she could identify several of the residents by name. My interlocutor had impressive expertise in civil engineering, had worked with building energy issues, and was additionally well-versed with energy poverty issues and policies.

The project was impressively well-rounded in its conceptualization and scope. Featuring a focus on circular economy, it included shared bicycles, repair cafés, shared vegetable garden plots, composting, and flea markets based on bartering, combining participatory activities with the built environment and renewable energy. For energy end-use, they had engaged residents using an app procured in partnership with a company to track energy usage, with 28 students auditing 28 households alongside residents. All of this was combined with advisory sessions and iterative engagement to build trust and awareness. Another round of building energy efficiency audits was scheduled for March 2022, involving 48 more students.

A representative from the company that had supplied the app talked about a 'cidadãos cientistas', a city of scientists, where everyone participated with a spirit of curiosity and interest in various themes. Residents would have a chance to reflect upon the findings presented this evening, and to continue engaging with project activities, towards deciding whether to establish a local REC with a solar plant on the public school. They would be able to do this with a highly detailed cost-benefit analysis down to household energy practices and infrastructure.

I was saddened to hear in July 2023, then, that the REC had not been enabled so far. One could hardly design a better configuration than the Smile Sintra project, in terms of a holistic approach to engagement and informed participation in energy transitions. Yet constrained timelines and resources do not often offer the luxury of such attempts, and it is a cause of concern that despite all this effort, the community did not proceed with an REC in a smooth manner. I was informed that one of the main local enthusiasts was put off upon learning that the size of the savings would not amount to more than a few Euro a month per household. This ran the risk of interest fading despite various corollary benefits integrated in the project design, with an option being to implement a collective self-consumption project that over time could evolve into a full-fledged energy community. The reasons are a matter for the project to analyse and discuss. My takeaway from this example is that despite proactive endeavours, RECs still had a long way to go in 2023 in Portugal.

The magic trick of a solar energy community

It is not all gloom and doom, however. With a healthy dose of realism, there were also some success stories, despite seemingly inevitable caveats, as in the Eurosolar4All Almada pilot. One such story comes from the Telheiras parish of Lisbon, with the Viver Telheiras REC, or Comunidade de Energia Renovável de Telheiras (CER Telheiras in Portuguese).[3] I first learnt about it in detail during a public event at the Lisbon Municipal Library on 17 February 2022, which brought together several experts on energy communities. The speakers emphasized that the incumbent energy company EDP's household solar products offered very limited benefits for households, whereas the new legislation enabling RECs constituted a pathway forward for communities to take charge of part of their own clean local energy production. This could be done in a manner that benefited the member households of an REC, and the experience of the Telheiras REC was important in this respect, well beyond its 7.4 kW capacity, 17 members or 11.5 MWh of annual production.

On 21 February 2022, I had a chance to interview one of the main forces behind this REC. The initiative had a strong commitment to solidarity, with some households taken along through co-funding arrangements despite being able to contribute to initial investment. As a local resident, my interlocutor could use his social network in the neighbourhood. But as a researcher, he could also draw on his expertise in multiple ways: "Being a researcher gives you legitimacy to know what you are talking about with energy policy." Partnering a local association had opened up doors, in-house expertise on energy audits and infrastructure had helped navigate complex processes towards getting REC certification. Occupying a dual role as an energy engineer with technical expertise on energy efficiency, and a social activist trying to mobilize decentralized energy models, made him see things in a cohesive way where others might struggle to fathom a connection. "One-stop shops can really help make change happen", he enthused, explaining how municipal energy agencies could help households access energy efficiency support, through guidance to navigate the complex bureaucratic requirements that support schemes were typically bundled up with.

We sat in one of my favourite spots in Lisbon, the Jardim do Príncipe Real, a park that embodies the possibility of commoning (see also Gato, 2016). Commoning is the antithesis of enclosure, making goods more widely available for all to use and value as stewards. I was touched by my interlocutor's earnest desire to do the same with local energy in his neighbourhood. He reflected on his evolving politics:

'I entered from a technical space, but focus increasingly on social aspects about implementation. Civil engineers are better than environmental engineers at many technical aspects, electrical engineers at solar, and so on. But the social interface is a useful middle ground. A lot of the social science theoretical literature on energy justice is interesting, but not very useful or easy to comprehend. Even your and [energy social scientist Benjamin] Sovacool's papers are sometimes hard to apply.'

This is a delightful aspect of conducting fieldwork, that it can shine a light on oneself, advancing self-reflection in entanglement with what one is trying to study. Stirling (2019) writes about this as the eagle eye versus worm eye view, questioning to what extent researchers are complicit in incumbency. My own engagement in energy social science has led me to increasingly reflect upon solidarity as what I call "our vessel of choice" in one of my occasional poems (Sareen, 2020, p 165), and in collaborative writing (Sareen et al, 2023a; 2023c).

My interlocutor had concrete ideas about things that needed changing, and soon:

> 'Solar rollout and whether it matters to people depends on the ownership, whether it benefits them; we are very far from it in Lisbon. But it is a matter of getting some projects in neighbourhoods, proof of concept, then it will work. Need to be fair returns, not four eurocents per kWh. … Now the government will launch a programme for small-scale solar with €30 million funding in PRR [the Recovery and Resilience Plan], with potential to catalyse smaller solar, energy communities. But it can also bring energy utilities to our roofs. But the current schemes attract people who can afford to invest €8,000 but would still like to get it back.'

To him, the challenge was clear term definition, efficient ways to proceed, and fair return on investment, to make the REC legislation come to life through operationalization in practice.[4] He envisaged football clubs, local associations, all embracing community solar. "I don't see it accelerating so much that we need to plan for limits to growth for energy communities", he said. "They seem to be planning ahead for this, not for anything else."

On 4 July 2023, I spent the afternoon observing a workshop on energy poverty, run in Portuguese by several researchers with a large variety of energy practitioners participating. Figure 6.2 shows a glimpse of this workshop. A presentation by a Viver Telheiras REC representative went step-by-step through their painstaking experience of establishing the REC. They were still waiting for approval, and it was only later that

Figure 6.2: A participatory workshop on energy poverty

month that it finally came through. This systematic approach to translating legislation into practice, to commoning the knowledge gained along the way (the presentation is also available via their website), is an important contribution towards accelerating the deployment of RECs in Portugal. It is a pity it had to come from this sort of patient bottom-up agency, whereas the boom in utility-scale solar was ushered in through a clearly defined top-down process by a state eager to attract capital from foreign investors.

★★★

Was it all a charade, then? Were the legislative changes to enable community energy more a public legitimation device than a manifestation of sincere intent backed by reasonable will to implement RECs? After all, the solar energy cooperative Coopérnico had been pushing for change for years, speaking at invited forums and giving inputs at public hearings to encourage more rapid enablement of solar energy cooperatives in citizen-centric forms. Why was it that legislation and public policy moved so slowly when it came to opening up the doors for community ownership of local clean energy sources, which would lighten people's energy poverty burden, reduce the need for the state to allocate large volumes of grid capacity, and see vast swathes of land taken over for solar generation?

My interviews with diverse stakeholders and multiple site visits had revealed a picture more complex than a binary yes/no answer would do justice to. I did not think the advisors at the Ministry of Environment and Energy Transition (subsequently changed to the Ministry of Environment and Climate Action) had any ill will towards the Portuguese people, nor that the state was keen to enrich foreign investors at the cost of impoverished households. I do think the handling of energy politics that led to the particular series of decisions and the prioritization across spatial scales was sub-optimal. Portugal missed a trick when it could have pushed to enable RECs in advance of the European directives that were known to be on their way. It put all its eggs in the utility-scale solar basket for all too long, and when the COVID-19 pandemic came along, it delayed implementation by a couple more years. Well-conceived initiatives on the ground faced a frustrating lack of response from the same state making grandiose announcements on energy transitions, and some saw prospective beneficiaries like household members of future RECs grow disillusioned. In the little places where individual agency worked to enable spatially distributed and collectively owned solar energy infrastructure, the state was a looming presence that offered only demands and delays, no encouragement, nor support.

It is telling that commercial intermediaries like CleanWatts and GreenVolt made far more rapid headway in upscaling their operating logic than the more participation oriented Coopérnico. These rapidly growing companies understood that within a neoliberal market logic of energy transition, a simple commercial offering that asked very little of households would have a better chance of gaining and retaining users. Their appeal was to the pocket: become a member of our local energy community, and save hundreds of Euro annually. It is not that Coopérnico lacked good ideas; they were remarkably successful in fundraising to install more solar capacity when they could find places that agreed to host solar plants. It is that the logic championed by the Portuguese government remained unsympathetic to cooperatives and collectives; it favoured market approaches that were a continuation of the commodified understanding of energy established during the fossil fuel era. Over the years, in many interviews with representatives of Coopérnico, this mismatch remained unmoved. Anyone keen to see evidence of protectionism in favour of the incumbent energy company has only to look at the great trials Coopérnico was put through, despite its laudable vision.[5]

★★★

I interviewed representatives of various energy associations over the years. They understood the perspectives of their members intimately, and some of them like the Portuguese Renewable Energy Association (APREN) were valuable, prominent voices in energy sector debates, also creating arenas for stakeholders to discuss changing industry dynamics and directions. On 12 July 2023, I had a chance to hold a repeat interview with an energy association representative; we had previously had a long discussion back on 23 July 2019. So much had changed, and yet so much had remained the same. He argued that the Ministry of Finance's unwillingness to invest in expanding human resources at DGEG was costing Portugal dearly, by withholding the benefits that multi-scalar solar deployment would bring in terms of productivity and equity. Empowering people would pay itself back many times over in their contribution to society and the improved wellbeing of people in a more locally self-sufficient country.

Lacking adequate human resources and thereby institutional capacity, authoritative public institutions, especially DGEG, had not processed proposals and were delaying the growth of solar energy, especially at lower scales. "Tightening public finance and delaying approval timelines is short-sighted, as it prevents value addition that would in turn benefit the Portuguese economy by generating more activity and taxes. Instead, the access to energy generation and revenue remains limited primarily to large companies."

The largest energy association, APREN, had incorporated small solar energy company associations, hence now also represented small solar growth due to consolidation among developers. There were considerable achievements to take pride in since 2017: solar sector growth from below 0.5 GW to over 2.5 GW; its clear prioritization in the national future; easy availability of foreign investment for Portuguese solar; and recognition of its competitiveness. Just in July 2023, the government more than doubled its solar target, from 9 GW by 2027 to 20.4 GW by 2030.

Yet, in this interview, the energy association representative voiced that this mature industry was frustrated by delays from national bureaucracy, the main thing that needed changing. "What people like you and me can contribute", he declared, "is to be vocal and increase recognition of what there is for Portugal to gain if more solar is built faster and at multiple scales to benefit various actors".

Various models were clear, he explained, in showing preferable pathways to adopt more renewable energy sources faster, with many savings and earnings, in line with Portuguese policies to electrify and decarbonize sectors. These pathways featured solar at all spatial scales. That week, there had been mention of a 0.4 per cent limit to land use for solar in Portugal, and I asked what he made of this. He confirmed my suspicion: there was no particular basis for this through rigorous analysis. It was a strategic pre-emptive measure by the government to contain discontent that they saw coming from the population about land being diverted to solar generation. By specifying a maximum limit, they could point to it and say it had been considered and dealt with, and that solar growth would not risk national interests.

He closed with this sobering thought:

'It is unlikely much small solar capacity will come online in time to meet a big part of the 2030 target [of 20.4 GW] as the timeline for increased energy flexibility would still leave very little room to reward generation during daily peak solar output periods, without storage, which is added by larger players at a higher scale of aggregation. So a lot of the absorption of a higher renewable energy mix will ride on large companies and their emerging business models, with greater integration of the Iberian electricity market.'

This resonated with my strong sense, as I draw out next in Chapter 7, that by 2023, a brief window of opportunity was rapidly closing for Portuguese community-scale solar energy. It was a good idea whose time had come. But protectionist regulation had held it back, so that by the time it could finally get going, utility-scale solar had already shaped the market trend.

SEVEN

Solar Portugal 2023

The need for cross-sectoral action writ large in renewed ambitions in 2023

When I stepped onto the stage in Carcavelos on 29 June 2023, the Portuguese government was soon to announce its bold upward revision of the national solar target, to 20.4 gigawatts (GW) by 2030. This would mark an eightfold increase within seven years. During the six-year span of 2017–2023, they had managed a fivefold increase, with less overall volume. This included the challenges of the COVID-19 pandemic circumstances, and of turning around a sector within the complex national political economy of energy, but it also relied on some low-hanging fruit – available transmission grid capacity, foreign investors eager to enter this emerging market, the buzz generated by the auctions of 2019 and 2020, and land availability in the spots that solar developers targeted. These were complex factors to maintain over time, especially land and grid access, let alone to keep accelerating with more ambition and larger volumes.

I dwelt on these aspects as I addressed the audience of nearly a hundred European delegates we had bussed in from Lisbon that morning, for a conference of the European Positive Energy

Districts Network, which I had had the privilege of helping to steer since 2020.[1] This audience included many key stakeholders helping make the vision of 100 Positive Energy Districts (hereafter PEDs) real by 2025. A PED was a neighbourhood block of buildings that produced more clean energy than it consumed. It represented a vision where distributed energy sources like solar had a major role to play alongside energy efficiency and energy flexibility solutions to change the spatial logics and temporal rhythms of energy systems. For years, I had worked collaboratively with colleagues in various parts of Europe and beyond to understand how the digitalization and decarbonization required to enable such transitions could take place in a socially inclusive way towards greater equity (Sareen et al, 2023b). Seeing this large audience full of people working to realize such ambitions was heartening. Yet here we were, in a large auditorium, with a set of relatively privileged people assembled for two days spent seeing best practices and success stories. The programme had little to do with the social struggle and vulnerability that I knew to be part of the everyday life of many Portuguese households, working hard to make do despite harsh odds. All too often, this is how even socially minded conferences about grand societal challenges play out – in posh settings with well-meaning people in a sort of utopian bubble.

Over the next two days, the group we had convened held many discussions about how to achieve the PED vision. We visited several sites of inspiration in both Lisbon and the Carcavelos parish of the tourism magnet that is Cascais municipality. Figure 7.1 shows conference participants gathered in Carcavelos. We saw examples of small distributed renewable energy plants, shared stories of frustration and innovative ways to resolve deadlocks, and reflected upon how these efforts could be transferred and replicated across many more contexts all over Europe. Cascais represented a suitable example of a relatively small municipality that had been innovative in acquiring large amounts of European funding to accelerate learning and piloting towards making PEDs a reality. The municipality was setting up a Renewable Energy

Figure 7.1: Positive Energy District conference participants out on a site visit in Carcavelos

Community (REC), and had attracted members who were keen to benefit from local solar energy production within a radius of several kilometres from the planned installation.

The municipality took a holistic view of sustainable urban development, and was investing in blue-green infrastructure to combat any future heat island effect from global warming. It was highly proactive in taking up challenges towards a low-carbon transition, and was trying various ways to share the benefits of a more sustainable lifestyle with residents, for instance offering local dwellers free public transport. While many people still had cars – necessitated by the legacy of sprawling spatial planning and suburbs that is a familiar story in many cities – urban development prioritized making it attractive to use active transport modes such as bicycles. One could poke holes and find flaws in how things were – our host noted several of these himself in the spirit of collegial sharing – but one could not fault the municipality for powering ahead towards a better future and a more equitable and low-carbon energy system. Its vision was infectious, and standing by the fabulous beach, looking out at the projects they had toured, conference participants were impressed and thinking about ways that PEDs could really come alive.

Conferences like this one play a valuable role in offering hope to people who have every reason to feel despondent, even disillusioned, by many of the trends that are painfully familiar to sustainability scientists and practitioners. So much change continues to go in the wrong direction: more highways are built, more forests are destroyed, more elite consumption options are prioritized in policies, more billionaires are given tax breaks, even as small-scale renewable energy development owned by local cooperatives is held back, modest public sector worker salaries decline relative to inflation, and poor households are unable to afford energy efficiency upgrades while facing rising monthly electricity costs. Finding people to share these frustrations and inspiration experiences with is important in order to keep working with dedication to visions of possible desirable futures full of PEDs.

Yet, when debriefing after two heady days with a colleague who had joined proceedings, we shared a sense in which all this had left us feeling somewhat empty. There had been almost no talk of the socioecological conflicts and costs of elite lifestyles that blazed on regardless, no platforming of discussions about the impact of extractive industries on landscapes where minerals crucial for Europe's energy transitions were being mined (often outside Europe). What we had missed was a sense of the political economic puzzle. There had also been little reflection on the people left out of any glamorous PED vision, such as the ones living in social housing as in the Eurosolar4All Almada pilot. If Cascais was a glowing beacon of how a municipality could land lots of sustainability funding and set numerous projects in play towards a brighter future, Almada municipality offered a sobering reminder that developments were contingent on many complex factors, and many a time these did not align in a virtuous direction. Social inclusion and equity, just energy transitions, these were notoriously slippery problems to grasp. Climate change was hardly the only wicked problem.

★★★

Our stellar host from Cascais municipality was all too aware of the nature of this challenge. On 4 March 2022, before Cascais had landed an impressive string of European projects and become a throbbing example of urban sustainability in action, he had been kind enough to take the time to show me around and explain the vision that he then hoped would unfold locally in the coming years. He had a plan, and considerable experience and networks to draw on in order to translate it rapidly into action. Already then, for instance, he could see that the recently announced legislation enabling PEDs would need practitioners to give feedback to help adapt it to emergent techno-economic models. He was far too well informed and a man of the world to be naïve about how things were likely to evolve:

'Two years ago on a farm, 700 hectares, they had a hunt, killed lots of animals. People filmed, it became a controversy, photos of hundreds of dead animals. Later, we found out the guy who owned the farm wanted the animals killed, so he could lease it to solar production. The city hall did not know, it was protected landscape so it was treated at the national level. And he got all the approvals. Social backlash, but it will proceed.'

This was the way of capital. Solar was the big game in town and many people wanted a slice of the pie. I had occasionally been approached by large landowners and brokers who had come across my research, and were keen to have discussions to understand whether a deal with solar developers made sense, and what to be aware of. I was careful about my level of engagement, willing to increase awareness but cautious not to become too personally involved in any development projects in a sector I was researching, and not to tilt any scales in favour of something I did not have sufficient information at hand to understand the fuller implications of. For my interlocutor, this sort of direct conflict between valuing nature and seeking rapid financial gain from large-scale solar deployment was best avoided by making other ways possible. His position allowed him to showcase other ways by implementation.

This vision was not limited to solar energy. Solar was a piece of the puzzle – a critical piece, but not one that could succeed in isolation. He had a firm grasp on the political economic picture, and thought in terms of energy systems rather than sectoral siloes like solar:

'EDP [Energias de Portugal] is the biggest green bonds buyer in the world, financed through centralized renewables production, they don't want decentralized solar to ruin their party. So we have to disrupt that model as the public sector. Our demographic in Portugal is not building many new buildings, it is retrofitting, refurbishing. So it is about adapting equipment to new reality, upgrading our buildings. So I can help make this happen in the right direction by making critical things mandatory. Then EDP will fall in line, get a little less on profits but help take our priorities forward. So my key role is to bring households on board.'

There it was, an activist vision, neatly packaged in political realism. Here was a doer with a plan. He thought the biggest bottleneck a municipality faced in a business model for local solar deployment was lack of clarity as an organization. Local decision-makers typically lacked the gumption to take on long-term debt as a municipality in order to deploy community-scale solar to serve their households. The REC market would take time to mature before building cooperatives could embrace it. This left frontrunners like Cascais in a situation of uncertainty, with a role to operationalize the emergent legislation on RECs into viable business models. Cascais was willing to invest in a 50 kilowatt (kW) solar plant on a public building rooftop, but did not have buy-in from enough members within a radius of 5 kilometres. This was the range admissible within legislation at the time.

'I have a school where we could put up 500 kW solar with self-consumption, it is connected to the medium-voltage grid.

And it could stream to households on the weekends, they just need to sign a contract', he declared. But the distribution system operator (DSO) did not trust the municipality, and he wondered what it would cost to use the grid to stream electricity locally within the REC. Would it be five eurocents per kW? That would kill the business case. Without clarity from the Energy Services Regulatory Authority (ERSE) on these crucial regulations, a municipality could not make a clear offer to households, could not promise, say, 50 kilowatt-hours per month at a given rate. He had also been surprised by how many people were unwilling to let others put up solar panels on the rooftops of their cooperatives. They were "afraid to lose a bit of control within the cooperative".

He had realized that all parties lacked knowledge on regulation:

'The electrical engineer is used to standard ways of running and installing a system. The building manager typically has no experience with new buildings with photovoltaics [PV]. Solar thermal has been mandatory, but heat pumps will move that need out in due course. The architects rule in Portugal, they push for flat roofs, not apt for this climate. They don't want new things. So there is this lack of embodied knowledge. My sister invested in a heat pump and her savings per year are €500. She will pay it off within five years.'

As a municipality with protected nature areas on one-third of its territory, Cascais had in-house capacity to undertake landscape reorientation, deal with invasive species.

"We have the capacity to do that in-house on biodiversity, that is our thing in Cascais. But we don't have that capacity on energy. I work on these electrical projects from a planning perspective. We know the capacity and possibilities with various public buildings." This meant they needed external know-how on energy, from private companies. But any private company being given the 'surface right of use' on a public building needed to go through a public tendering process, and to fulfil criteria on quality of service during delivery. Developers were unwilling on take on this liability, and investors were turned off by the worry that legal issues would cause delays. This was so much less attractive for them when they could develop utility-scale solar with attractive power purchase agreements up for grabs with companies.

My interlocutor thus had a mountain to climb. Cascais had a 185 megawatt (MW) solar capacity target by 2030. By 2022, the municipality itself had only installed 4 MW solar capacity. But a private investor had installed two more, with licences through the Directorate General of Energy and Geology (DGEG) followed by approval from the DSO. The municipal strategy was to attract external project funds to experiment and ensure compliance with the new regulations on RECs, while aiming to install 500 kW community-scale solar on buildings such as the local university campus, to feed to local households as REC members. If such a model took off, hopefully this would pave the way to convince others to welcome public solar installations on their rooftops and lands based on a leasing model, and to join other local RECs. The municipal worker was savvy about the need to increase public awareness for this to work over time. There were plans to hire staff to give advisory services as a one-stop shop on RECs, something that securing European development projects could enable. It was thus that Cascais would bring a PED vision to life.

While critical of aspects of the energy market, my interlocutor had plans that could conceivably work within it: "Let's say an investor comes in with interest in RECs. We need to have terms of reference which convince him. The city will use all the surplus energy if the households don't. So he knows he will get his fixed revenue flow." The city could drive safe investment, coordinating where solar capacity came up along with electric vehicle charging infrastructure, co-locating these through its spatial planning. He understood the priority of the DSO as being antithetical to RECs. E-Redes, he said, wanted:

'80 per cent of their revenue in 2030 to come from services, not production. The smart meters they installed are not two-way. This is not coincidental. They want to make it difficult for citizens to change. Two-way meters – you pay more to rent it than one-way meters, so people do not have a value proposition. The installation commissions ruin the household value proposition.'

Here it was again, clear recognition of the real value being lost for the Portuguese people by the regulator and ministry's unwillingness to value the electricity distribution grid in tune with the reality of a digitalized energy system that would electrify multiple sectors to enable a low-carbon transition. The ship had sailed, ERSE had greenlighted the rapid deployment of smart electric meters to all Portuguese households by the DSO in a way that allowed this critical energy infrastructure to be shaped in line with the incumbent's self-interest, at the cost of democratizing control and ownership over energy systems. Even if the municipalities or another DSO took over the licence concessions from E-Redes at some future point, E-Redes would recover its investments handsomely, and the grid would remain aligned to centralized control, resisting high remunerative value to local solar energy generation.

A municipal worker tuned into the peculiar version of the energy transition being allowed through incumbency politics in Portugal could understand all this, and battle for local households' future energy sector stake by trying to establish RECs against formidable odds. Conferences showcasing even the modest achievements possible despite this travesty of justice – like our PED conference on 29–30 June 2023 – were counter-hegemonic in relation to this bigger picture. They did not challenge the overall energy market model outright – a pragmatic view did not suggest much hope of success lying in that direction. The framing of such conferences is to some extent co-opted by the hegemonic outlook on energy – as a commodity, as an entrenched sector with centrally set politics and policies that determine systemic control. But within this nationally – and globally – dominant framing, they nonetheless insert rays of hope and visions of possibilities that could salvage some local wins without disrupting the energy system entirely. Conferences of this sort insert the kernel of counter-hegemonic possibilities towards limited forms of energy democracy into the minds of decision-makers and high-level practitioners who would tune out entirely from something that railed against the powers that be and dismissed the authoritative institutions that structure their everyday practices of energy governance.

The politics my interlocutor in Cascais articulated in our pleasant and insightful morning together in 2022, and then performed publicly when hosting the conference in 2023, were the plucky politics of negotiation. He was navigating between a rock and a hard place. This was not a refusal to acknowledge the political economy of energy. It was part of a dance within that political economy, born of the recognition that this was the best hope at this moment in time to work towards more limited forms of change from the inside. In all likelihood, he preferred this rather than to be left out in the cold, dreaming of a vision of energy transition that had failed to enrol the Portuguese politicians and administrators. In the absence of strong accountability mechanisms towards a truly just energy transition, and faced with the need to enable rapid action, what else could a person do?

★★★

On 27 July 2023, three weeks after receiving the latest response in my long-term correspondence with DGEG officers, and a fortnight after having concluded my fieldwork of 2023 in Portugal, I took them up on their offer to respond to emailed questions. My reflections from the past month of interviews and observations on evolving Portuguese solar energy governance and prospects of just energy transitions had crystallized into some questions that it made sense to put to this agency that ran the Portuguese energy sector. These are the five questions I emailed them:

1. During my research on this topic, both since 2017 and now during June–July 2023 in Lisbon, many energy sector stakeholders have expressed considerable frustration that DGEG has been unable to deal with applications efficiently in a time-bound manner. In particular this concerns collective self-consumption and renewable energy community applications, which I almost invariably hear are waitlisted for months. *What are the reasons DGEG has faced human resource constraints in processing applications for collective self-consumption and renewable energy communities, and what actions are being taken to resolve this problem and backlog?*

2. I have been involved in the social impact assessment of one such project, by AGENEAL in Almada municipality, who were recently (this week!) awarded authorization of their collective self-consumption project to provide benefits to families in social housing. This approval took a long time. *Does DGEG expect that future approvals for collective self-consumption projects, especially those serving lower income groups, will be expedited and handled with shorter timelines? What is your target timeline from the date of application and by when do you expect to achieve this routinely in practice?*

3. In expert interviews during my research, I have been informed that very few renewable energy community applications have been approved (only three as of June 2023) and also very few applications for collective self-consumption (waitlist of over 300 as of June 2023). *Are these numbers correct, and can you share updated numbers for both? For the beneficiaries, what is the difference in practice of being approved under these different forms of recognition?*

4. I am aware that companies like Cleanvolt and Greenwatts are proliferating new models rapidly, and have had some success in earning money this way as well as sharing some benefits with households. At the same time, companies are also

applying for approvals as renewable energy communities. *What is DGEG's position on these emerging dynamics? Do you have a preferred pathway of growth for renewable energy communities from your institutional perspective, in terms of what is desirable for the Portuguese renewable energy and electricity sectors? If so, how do you promote this preference?*

5. I was surprised to be informed that the low voltage grid concessions have still not been awarded for a future long-term period, as this matter has been pending for several years now (at least since 2019). One of the issues emphasized during parliamentary hearings and elsewhere (including in my research) is the importance of a valuation of the distribution grid, bearing in mind the renewed importance of a smart grid in a flexibility market for electricity, which is fast emerging in Portugal as the penetration of renewable energy has increased in the grid mix. *Is such a valuation of the digitalized grid being done? Do you along with regulator ERSE have an envisioned timeline for the award of grid concessions? If not, what sub-optimal results do you foresee due to this for the proliferation of renewable energy communities and collective self-consumption, given that these require up-to-date grid infrastructure? How do you handle the complexity with having the current DSO invest in distribution grid upgrades?*

I received a response within a matter of hours the same day, from a high-ranking official. She informed and assured me that: 'The matter in question is within the area of competence of DGEG's electricity department. In this sense, we forward the questions to the respective department, and we are waiting for their response. As soon as possible we will send you the information.' More than three months later, I had still not received any answer to my five questions. Over the years, I had interviewed senior officials at DGEG and ERSE, and found them forthcoming – if also politically correct, mindful of their important societal roles and responsibilities – and willing to engage despite their overburdened schedules. Unlike some of my interviewees who had mentioned dogged persistence in trying to get answers out of DGEG, I did not think it was my place to be pushy in this instance. As a prominent researcher of solar energy governance in Portugal, having contributed insights on the sector from my base abroad over the years in a constructive, respectful way, I wanted to see if this national agency would show me the courtesy of at least attempting an answer to my blunt but clearly well-informed and pertinent questions. Their failure to prioritize even providing a placeholder response is telling in itself, and of a piece with many interlocutors' frustrating experience of having to navigate endless delay and silence.

I include my questions here both because they merit answers, and in the hope that readers will put them to DGEG themselves. Prefigurative politics are important: if research-based insights can not only shine a light on systemic gaps, but inform interested publics about what gaps to poke at, then people can exercise individual and collective agency and organize around this. Beyond rooms of experts, there is little engagement with decision-making in a sector still perceived as distant, technical and top-down by Portuguese residents. A growing part of the population raising these questions in righteous indignation would have a very different effect on DGEG, ERSE, the Secretary of State for Energy and the Ministry of Climate and Energy Transition. It would challenge the legitimacy of such energy transitions governance, and push the envelope to strengthen bottom-up accountability.

A past that is present in the future

Community solar fought its bureaucratic battles, but rapid movement on utility-scale solar continued in 2023. While the last coal thermal plant in Portugal was closed in 2021 in Sines, there were energy security concerns due to the Russian war in Ukraine and high electricity prices in Europe in 2022 and 2023, which meant that it remained available on standby. Even so, it was clear that the transmission grid had 1.2 GW capacity that solar could fill near this location, and Spanish energy major Iberdrola moved to build 'Fernando Pessoa', Europe's largest solar plant in Santiago do Cacém, with renewable energy company Prosolia Solar. Environmental clearance came through in 2023, targeting a commissioning date in 2025.

On 7 July 2023, a collaborator from a Portuguese university and I drove to this location in the Alentejo region. Media had reported that a plantation of 1.5 million eucalyptus trees would be removed to make way for this plant. The chopping down of over 1,800 cork oak trees, heavily protected under Portuguese regulations, had also been approved by the Portuguese Environment Agency in this exceptional case. Along the way, we made stops in the parishes of Sao Domingos and Vale de Agua, near which other solar plants had been proposed around Cercal. These villages lay quiet with gleaming pavements and tiled walls in the summer sun, and with hardly a soul about. We held a long discussion with an official, confirming what we had surmised through visits to other small municipalities and parish offices. While there had been some unease and discussion with regard to solar development, to many it was simply a matter that solar developers dealt with along with those local stakeholders from whom they leased the land, and with those greenfield developers and intermediaries involved in securing or issuing environmental and regulatory clearances. An hour or two out of Lisbon in cork-studded landscapes, there was little public involvement in Portugal's solar energy transition, little evidence of solar as unlocking energy democracy or creating a just energy system. These were not large or powerful municipalities proximate to the national capital or big tourist draws, they were small towns sprinkled across this poor and sparsely populated region with an ageing population, and soon with fewer cork oaks.

We did come across posters with information about a public hearing about a solar plant that had been proposed and had encountered local opposition. I subsequently heard from actors who had been involved in this project in an advisory role. They noted the developer had not taken the importance of local consultations seriously enough, which had led to pushback by indignant local residents who felt shunted. They disapproved of an attitude of outsiders coming in and expecting to simply get their way without discussion, and did not agree to the solar development. So public involvement was not altogether missing, either. Residents did expect recognition of and respect for formal and customary ways in which local projects could proceed.

After some triangulation using limited information available online and verbal guidance from people we asked along the way, and satellite views on Google Maps, we found the massive commercial eucalyptus plantation that would soon be replaced by sprawling arrays of solar panels. We walked the length and part of the breadth of this massive expanse, with spiky tall trees that varied in height and girth based on when the rows had been planted. It was a remarkable experience, knowing that this landscape – heavily shaped by humans to begin with – would soon change beyond recognition. Figure 7.2 gives an impression of this plantation, whose scope cannot be

Figure 7.2: The 1.5 million tree eucalyptus plantation that the 1.2 gigawatt solar plant will replace

Figure 7.3: Mechanized felling of a eucalyptus plantation

captured in one image. Along the way, we came by another smaller eucalyptus plantation, one that was being cleared that very day. As we stopped to watch, the tall trees were lopped down before our eyes, by one person with one machine, while a couple of his colleagues looked on. Figure 7.3 shows this eucalyptus plantation being felled. The speed with which industrial interventions can alter a landscape is startling. We reflected upon how it would not take long for all traces of the cork oaks and eucalyptus and other trees and shrubs that the solar panels replaced to disappear from social memory altogether, as the hapless champions of the Via Algarviana had learnt when the 220 MW Alcoutim solar plant had taken over part of that landscape.

Utility-scale solar has already taken over some rural Portuguese landscapes. Driving northwest to the closed coal thermal plant in Sines, we stopped at a 44 MW solar park. Figure 7.4 shows this large solar park, with the chimneys of Sines looming in the background. For a while, we walked along the fence alongside it. I was struck by its fenced aloofness as a silent, mighty producer. A month prior to this, in June 2023, I had looked at the largest solar plant in Uganda, the 20 MW Kabulasoke. Later in July 2023, I would visit some

Figure 7.4: Large solar plant with the Sines coal thermal plant in
the background

of the GW-scale solar plants in the Indian state of Rajasthan.
The transmission grids being rapidly developed in its desert
reaches were mindboggling in both their proportion and
speed of implementation. The Power Grid Corporation had
put up 765 kilovolt transmission lines, the highest permissible
capacity in India, to evacuate these GW of solar power out
of Rajasthan to energy demand centres in places such as the
Northern Capital Region around New Delhi.

Portugal was not alone in reproducing the spatial logic of
fossil fuels with solar energy, as it sped up efforts to reach its
ambitious climate targets. In doing so, like increasingly many
other countries, it missed the rare opportunity of utilizing the
modular flexibility of solar energy to put in place a spatially
distributed, more efficient network of energy generation
that lent itself to collective ownership and consumption in
nearby centres of energy demand. This was not incidental. It
was a manifestation of the way that capital works in a highly
financialized sector like renewable energy. Large energy
companies and foreign investors were able to shape the national
agenda by convincing governments that they could only
succeed in their mission to bring about an energy transition

if they aligned this ambition to the preferences of the large capitalist beneficiaries of such an energy transition. Promising their populace that cheap solar energy would lighten their financial burdens, and hyping up the record-low solar auction prices in places like Rajasthan and Portugal, these governments set the wheels in motion that let private companies corner sectoral growth. Low-carbon content increased on their electricity grids, landscapes changed their socioecological value and functions, and the hinterland became a place that produced power to feed cities and industries, with profits going to some company coffers elsewhere. Though energy sector debt did ease up for the government, residents neither felt lower electricity tariffs on the retail market, nor had a chance to invest in a means of low-carbon energy production that so easily and intuitively lent itself to collective ownership forms.

EIGHT

Conclusion

Lessons for policy and research for just solar energy transitions

The world is an increasingly connected place. My research on Portuguese solar governance has led to invitations to give talks regarding larger sectoral trends and needs at a global level. On two such occasions, I addressed audiences within the United Nations system. The first time, this was as part of the United Nations High Level Political Forum on 11 July 2018, at the Economic and Social Council in New York. I had contributed to policy briefs guiding the review of the Sustainable Development Goal (SDG) 7 on ensuring access to affordable, reliable, sustainable and modern energy for all. The Norwegian Ministry of Foreign Affairs and the SDG 7 Technical Advisory Group recognized the relevance of my inputs, and invited me to contribute to a panel discussion on the interlinkages between SDG 7 and other SDGs. This is a transcript of my speech at this forum:

> We've heard the good news, that there is a technological and economic argument for clean energy transitions, and it is working. But the way these transitions happen is based on

political economic drivers. We did a 15 state comparative study of electricity distribution in India. It's a project called Mapping Power. We found that political decision-making leads to vastly different outcomes under similar technical conditions. So it's clear that energy transitions won't lead to sustainability by themselves. Environmental benefits, yes, if we ensure sustainable production and consumption, that's SDG 12. But benefits to labour, to consumers, to citizens? That requires us to address energy vulnerabilities of marginalized groups. Issues of energy justice, energy poverty. We have a 30 country initiative in Europe, a COST Action called ENGAGER. We have put energy poverty on the map and worked to institute reporting mandates at the country level. That means developing energy poverty indicators that don't exist yet, putting the issue on the agenda for action.

Which brings me to my final, perhaps most important point. Indicators matter, but they alone will not achieve SDG 7 unless we understand that indicators don't just measure progress. They help direct change in the energy sector, they change its nature. So we need to stay nimble, as we are doing with the first review of SDG 7. But also to keep in mind that some things will always be hard to capture through the best of indicators – 'full' energy access, thermal comfort in people's experiences, infrastructure support moved away from fossil fuels towards renewable energy. We heard a lot about Portugal's energy transition at the review yesterday. We have just released a study showing that this transition must address both socio-technical and justice issues together to be sustainable. Regulatory foresight must be responsive to the public interest, and social science research on energy is important to achieve this understanding and overcome sectoral inertia. As I see it, the best way to think about the challenge of realizing SDG 7 is in terms of accountability – how do we institutionalize self-correcting systems towards equity? Indicators are a useful means, but this is the actual task to reach SDG 7.

Nearly five years to the day, I was invited to address the United Nations Conference on Trade and Development's Multi-year Expert Meeting on Trade, Services and Development, on 10 July 2023 in Geneva. Prioritizing fieldwork in Portugal, I joined remotely this time, having the privilege of being part of a panel that included a former chairman of the Energy Services Regulatory Authority (ERSE). The three-day meeting focused on 'the role of trade and services for enhancing science, technology and innovation to promote a fair transition to sustainable energy'. We were asked to share our experiences with policy making in relation to trade in services in support of energy transitions. I gave this speech with the intent of emphasizing three key points:

> Energy transitions rely on multiple sectors, at present chiefly electricity, transport, the built environment, and a variety of industries. Digitalization – not just the transition from analogue to digital but also its wider effects in socioeconomic and political spheres – is enabling the real-time coordination of activities across sectors. This coordination is vital for electrification of all sectors and thereby decarbonization, as electricity is relatively simple to supply using renewable energy sources, which are economically competitive.
>
> As sectors like electricity, transport and the built environment are digitalized and decarbonized, a variety of challenges manifest, cutting across sectors. One major challenge is social inclusion – many actors such as the elderly and less formally educated can be systemically excluded during digitalization of sectors, or otherwise put at a disadvantage. For instance, those paying bus fares manually usually pay more than those using a smart card or smartphone. Another major challenge is getting automated energy efficiency measures to benefit people, and not just supply side actors. Take the example of smart electric meters, which have largely benefitted incumbent

electricity utilities, and not led to fair sharing of those lower costs with users, despite such good intents and promises.

Making decarbonization and digitalization – or the twin transition – fair is important to achieve just energy transitions. Research has identified three key solutions in this regard: (1) transition metrics must be cross-sectoral, as digitalization will combine performance across sectors, for example through energy flexibility; (2) users must be treated as energy citizens, exercising both rights and responsibilities, and therefore involved throughout decision-making processes to express and enforce user priorities; and (3) the benefits of energy transitions must be shared in an equitable manner backed by sanctions to supply side actors if this fails, to achieve energy poverty alleviation, greater public legitimacy, and increased public ownership of and involvement in energy infrastructure and governance.

Revisiting these inputs, I am able to stand by them with a clear conscience. These are the main messages from my research on the governance of solar energy in Portugal that I want policy makers to heed across levels of governance. Even talks at prominent global platforms have limited effect if they only reach the experts in the room. Placing these texts here for posterity is an attempt to exercise my agency as an energy social scientist for social change.

★★★

To juxtapose these grand settings with something humbler, I turn to two stark memories from fieldwork over the years. The first is of my first fortnight in Portugal, spent in the Tamera eco-community in September 2017.[1] The second is a conversation with two of Portugal's earliest rooftop solar adopters, chatting at the University of Lisbon on 11 July 2023, the last week I spent in the country that year.

In Tamera, I learnt about how this community of about 200 people had worked to develop community-scale solar. For years, they had been running a kitchen that served 50 people a day, on a combination of solar energy as well as biomass from the 4,000 visitors the eco-community received annually. Solar energy was truly used as a convivial tool in this setting, a technology that helped people thrive. It was used in context-appropriate configurations, combining the use of multiple types of solar, such as solar thermal and concentrating solar power. Solar photovoltaic (PV) was used as well, but not in cooking; rather, it helped to power basic electricity needs as a modest mini-grid. The construction materials used in the solar kitchen and linked dining space had low embodied energy, meaning their use had not caused significant greenhouse gas emissions. The built environment was also shaped for the climate, with good air flow and a sense of coolness despite the hot temperatures for large parts of the year.

I could imagine it becoming a bit chilly during the short but cold winter, and while I did visit Tamera again in August 2018, I never experienced the solar kitchen during winter, when the eco-community is not open to visitors. During my first stay, however, I did help to build a seasonal system for space heating. Based on calculations by two engineers visiting from the United Kingdom, and inputs of a variety of locally available materials ranging from rope to urine (which I helped lug by the tens of litres in jerricans), we experimented with the use of exothermic composting processes as a winter heat source for an off-grid building. I was later informed that this failed utterly, but it is an example of the sort of frugal innovation that was part of the community's approach to living in a responsible way in their setting within limited means.

In keeping with this experimental attitude, two physicists had spent years working to develop concentrating solar power using Fresnel lenses in a giant disc configuration, to concentrate solar heat at a focal point for flexible purposes. Their vision included using it to heat a stone that could retain heat for a long time, and to attaining sufficiently high temperatures to power a cottage industry kiln to manufacture durable building materials locally. Heating a stone this way allowed making pizza on it after sunset, so that this could be served hot at a natural dinner time for their everyday rhythm. The kiln idea was a work in progress, but the vision behind these applications was to use solar energy in plural and customized ways. So different from the homogenizing and monocultural vision of utility-scale solar plants!

Tamera had also worked on integrating concentrating solar power with agricultural processes, using it to heat vegetable oils that could then optimize greenhouse performance to cultivate food throughout the year. The possibilities were endless, and their intent was to explore what could function, as a precursor to helping scale it up well beyond Tamera, by inspiring the visitors and using the attention that the eco-community attracted through experimentation and advocacy. This idea of cross-sectoral integration cuts across all these years and my growing conviction that this is an absolute necessity for just energy transitions further into the 2020s. It also draws a direct line to my conversation with two of Portugal's earliest solar adopters, as we sat together at the University of Lisbon on a July 2023 afternoon.

The University of Lisbon had recently installed an agrivoltaics facility, shown in Figure 8.1. I had noticed it on my way to this meeting. What on previous visits had been a plot reserved for a weather monitoring station and some solar panels, now had solar PV mounted on frames high above the ground, with space for light to pass through, and crop cultivation below. As a term, agrivoltaics comes from the merger of agriculture and photovoltaics, and here it was. Yet, in our conversation, the notes we struck were anything but optimistic.

These were senior researchers with decades of experience and an overarching understanding of energy systems, spanning not only solar energy but also electricity grids and transport electrification. I had interviewed and learnt a good deal from one of them over the years. On 6 March 2019, he had commented that cases like Tamera "are not scalable, they

Figure 8.1: Agrivoltaics facility installed at the University of Lisbon

depend on specific contexts. So there is a need for fungible models for community solar energy. Automation is clearly part of the way forward". In another interview on 30 July 2019, shortly after regulatory changes to enable renewable energy communities had been announced, he had stated, quite prophetically:

> 'I have not seen the small-scale draft, but expect it leaves things open without specifying a procedure for revenue-sharing, so from the academic point of view it is good as it allows experimentation without closing out options. There is no clear way of doing things properly yet. There will probably be some pilots driven by people within some buildings, top-down projects by city councils, and so on. So this would set the framework for future regulations or procedures.'

He was also mindful of a Chinese stake in ownership of Portuguese energy infrastructure. While this had been controversial in the country, it had been the way out of energy sector debt for an economy in the doldrums. But what intrigued his analytically sharp mind was the clear trend emerging from China that battery charging for electric vehicles would combine with electricity grid flexibility to play an increasingly significant role as solar energy penetration increased rapidly. China was not only the global leader in these respects, it was also a manufacturing superpower keen to lead on building energy infrastructure in order to determine global industry standards and consolidate upon its frontrunner position.

So when we had our discussion on 11 July 2023, I was curious as to his reading of recent trends. Both these interlocutors had their heart set on small-scale solar as the obvious desirable course for prioritized development, but they were too rational and analytically inclined to let this cloud their perspective on which way things were headed. When incentives for household rooftop solar had existed, over a decade ago, they had been among the privileged early adopters who could cash in on these, but were aware that households with lower incomes could not have benefited in the same way, unable to make the up-front investment. Now, having seen how several solar energy community pilots they were aware of had fared, they had surmised that these faced too many bureaucratic hurdles and also lacked enough "real community".

Smile Sintra was an instructive example in this sense, having faced considerable challenges to get buy-in from residents to become members of a Renewable Energy Community, because they did not find the modest savings enough of an incentive to be easily convinced about the value of collective engagement with the bureaucratic process. Moreover, to work, solar energy communities needed to scale up rapidly through replication of the operating logic, and this had simply not happened. They ran through examples. In their view, Coopérnico had been limited by its keenness to be involved everywhere instead of spawning other similar cooperatives, while the Viver Telheiras REC was rare and dependent on a motivated person as the main driver. Although a commercial intermediary like CleanWatts could enable community-scale collective self-consumption, it shared limited benefits with participating entities (the company itself took a substantial cut). Moreover, they expected its rollout would hardly compare with utility-scale solar, even if it did manage rapid diffusion. By 2030, they argued, so much large-scale solar would come on-grid before enough of the distributed solar capacity, that it would hollow out pricing during peak production hours, and flexibility would be met with smart charging of vehicles.

These were shrewd conclusions drawn from known trends. We discussed a general societal reluctance to engage in collectives and cooperatives, partly due to a negative association with communism, and due to a Portuguese building boom in the 1970s from which many housing cooperatives ran into financial trouble in the 1980s. They had contrasting takes on whether cooperatives were making a comeback in the 2020s. But they were excited about battery swapping models for transport. This to them presented an obvious, optimal way to make use of solar capacity and flexibility to decarbonize multi-range transport, while also over-building solar energy capacity that could feed battery charging whenever supply exceeded demand on the grid, instead of being curtailed. Emergent research suggested weight behind such strategic sectoral coupling plans (Vallera et al, 2021).

This realism came from experts who nonetheless continued to promote small-scale solar, something the agrivoltaics plant and a rooftop solar plant on the university campus were evidence of. Their conviction that small-scale solar would remain marginal in the big picture going forward was well-informed. Addressing the increased solar target of 20.4 gigawatts (GW) by 2030, they remarked that this was not so surprising. What had raised eyebrows in the announcement a week ago were large ambitions tied to green hydrogen and offshore wind, which they said nobody believed, knowing that these were signals to attract foreign investors, who themselves were unlikely to take such grandiose targets seriously.

The big picture on Portuguese solar had been settled for some time to come from the government's perspective – there was interest among foreign investors, and capacity was being rapidly installed. This was enough to label it a success in terms of reductive target-setting; this major upward revision would continue to spur growth. The priorities of the state had already moved on to greener pastures that could unlock further investment prospects. Even as community solar proponents waited for answers and approvals, the national agencies were doubtless busy with mobilizing these new priorities that made headlines. The trick the government had missed, in my view, was thinking not only in terms of GW installed, but also in terms of the spatial scales of installation, the ownership configurations and benefit flows, and the importance of cross-sectoral integration for energy flexibility on an electricity grid with rapidly increasing renewable energy content. This would prove vital as sectors such as transport were electrified and brought new demand patterns and possibilities into the mix.

The slow vision of Solis or an idea whose time has come?

Soon after my visit to the agrivoltaics facility, I noticed that Solis was organizing a study visit there for enthusiasts. Solis was the solar promotion initiative of Lisbon municipality, through its municipal energy agency, Lisboa Enova.[2] From Lisbon's vision of becoming a solar city, articulated in advance of its European Green Capital 2020 award, this initiative had emerged and grown steadily over the years. I held several interviews with Lisboa Enova representatives over the years. While Solis had not really taken off in terms of going viral on social media or getting heavy downloads of its mobile phone application, the agency kept building it up steadily. They gamified the process of mapping all the small-scale solar capacity that had been installed within Lisbon's boundaries. Reports could be crowd-sourced. In my months in Lisbon over the years, I observed a noticeable increase in urban solar, with more of it appearing on rooftops visible from the hilly city's many *miradouros*.

When I visited the municipal energy agency on 3 March 2022, their vision was to deploy 100 megawatts (MW) of small-scale solar in Lisbon by 2030. Less than 185 MW in Cascais, but ambitious given the far greater density of the capital. They had used a digital marketing team, but lacked a budget to hire social media influencers. Their ambition was for the Solis platform to attract and allow for diverse forms of engagement to map and drive solar deployment. Ideally, it would grow beyond Lisbon into a tool that other cities could use as well. They drew on learning from various research and development projects. One had featured not only 15 kilowatts (kW) of rooftop solar, but its integration to feed electric vehicle charging and the building's heating, ventilation and air cooling system. Larger systems would behave differently, but technology would only become more space-efficient and the business case would continue to improve.

A plan to install a 2 MW PV system in Carnide to inject direct current into one end of the dedicated subway grid had faced many setbacks due to a complex public tendering process. First there was a lack of interest, then a winning bid was successfully challenged on parametric compliance by competitors. Political changes in 2021 slowed things down, and meanwhile, costs went up, with the tender being put off. Then it turned into a green hydrogen idea, linked to future bus procurement, caught up in political dilly-dallying. Another prestige project, a rooftop solar plant on the city hall, was:

> '[A] nightmare when dealing with the national directorate for cultural heritage. They imposed lots of problems and constraints, so that the panels would not be visible. But we had some architectural inputs and devised a way to minimize aesthetic impact, but the approval took

two years. Now we are helping another institution that does a lot of interventions in public buildings in the municipality, they are aiming to renovate 20 buildings now. One is near the city hall, and they almost decided to skip PV systems because they did not want to deal with bureaucratic hassle.'

Thus, an important outcome of having a consistent solar priority at the municipal level, with dedicated resources and follow-up, was that it revealed forms of resistance and bureaucracy due to institutional and infrastructural obduracy. Projects were routinely postponed, but this was not the same as failure; it identified things that needed to change for systemic enablement of small-scale urban solar. Figure 8.2 shows a 20 kW solar plant installed a decade ago in Lisbon, now lying dusty, inefficient and neglected on a restaurant roof. Solar repair and maintenance were themes that required attention at the small scale, too.

One of them mused: "But why are we struggling to implement solar in areas with so many constraints, instead of targeting areas where it is easy? With energy communities, it

Figure 8.2: Dusty inefficient solar plant on a Lisbon restaurant

would be easier to take it forward in areas where it is more straight-forward." The other chimed in:

'There are some attempts at RECs, but none fit the definition yet, they are not operational. In the first two years since the new legislation, there have been a lot of discussions to understand how to implement RECs, and even the regulator ERSE has been learning. So ERSE has been willing to give regulatory sandboxing approvals for various pilot projects. So we have been channeling our work on these pilots and convincing the municipality to take this direction, to use excess production for social housing units. [A colleague I know] likes to call it a "solar tariff" for subsidized electricity this way. There is political will for this, it is part of governance goals for the four years until 2025.'

They had followed up on the European Green Capital with an in-house initiative called the 'compromiso verde' (green commitment). Two hundred companies had signed a pact with them, of which over 70 committed to install solar; but many of these already had it in place and some were outside city boundaries. They were planning green tours, like the one in 2023 that took enthusiasts to see the agrivoltaics facility. They had visited a high-end solar development with a sophisticated energy management model that used smart charging for energy flexibility to maximize solar self-consumption. But they wanted something simple to install everywhere, not just for rich real estate.

This reminded me of an interview with one of the most senior and respected figures in solar energy in Portugal just two days earlier, on 1 March 2022. I had met him on several occasions since 2017, and in this interview he mentioned advising a leading real estate firm on "cutting-edge sustainability measures as part of the branding for landmark projects. Money is not an issue so it allows me to implement all my energy efficiency dreams". The irony of doing this for a development in Comporta, "the Hamptons of Portugal", was not lost on him, but he justified this as creating innovation others could emulate cheaply. Yet here were my interlocutors at Lisboa Enova, saying that: "There is no private-public partnership model where we take the innovation of these elites to increase penetration in lower-income segments." The other chipped in: "I don't see demand response coming to Portugal with remote shutting down of equipment for flexibility based on automated incentives. People like their freedom." And the first agreed: "Let's not have AI [artificial intelligence] ruling our lives, that is a general sense. Maybe in the premium sector, because elites really like to buy fancy technology and show off to friends." They did, however, expect to see aggregation in flexibility services, with a revenue-sharing arrangement for households.

Moreover, they saw value in partnering with popular associations, like football clubs. Greenvolt was installing solar around the Os Belenenses football stadium to make an REC. Importantly, they had invested strategic effort to build a knowledge base, by surveying a representative sample of 1,500 respondents from Lisbon's parishes on public health, energy efficiency and building characteristics to map multiple dimensions of energy poverty.

We met again on 13 July 2023, my last interview before I left Portugal that year. Progress had been slow and steady. They were doing outreach with schools, working to integrate solar awareness in educational initiatives; inherently efforts with a long-term orientation. Unlike my interlocutors at the University of Lisbon, they were ecstatic about the Viver Telheiras REC and saw it as highly replicable. They admitted that inadequate human resource allocation to public sector institutions such as the Directorate General of Energy and Geology (DGEG) was holding things back. But they were hopeful that with an example in place in the Telheiras parish, others would replicate it. For them, the distribution system operator (DSO) had been frustrating to work with, as it still did not offer a platform to consistently retrieve smart meter data on energy consumption with 15-minute pulses:

'For years we asked them for an API [application programming interface] that would help us to retrieve this consistently every time they made changes to their websites. It is only now that E-Redes is almost ready to offer such a platform. This is the basis to run energy communities. And this simple development has taken three years. Even though there is work to transpose legislation, three years is still a lot. Digitalization is crucial in this transition.'

When it came to radius limitations to include members in an REC, they emphasized that the National Energy Agency (ADENE) had said it would consider all buildings connected to one sub-station eligible. But they feared this would be hard to verify: "It will go to DGEG or the DSO to be checked, and then if they approve it, project participants can proceed." Provided they could get through all this red tape in due course, they saw scope to couple urban solar with electric vehicle charging points, although this would need an increase in contracted power for building blocks, to ensure grid robustness. These were big ifs, with a lot riding on national actors who had not shown a tendency to prioritize rapid action in this direction.

They admitted their limitations in giving technical inputs on how to implement RECs: "[T]here are no easy answers even for us to give either. ADENE will publish such a guide. It is hard for us even when we break it down into small steps, because queries come up, and the only way to know is learning by doing, like Telheiras." We left off on a poignant question one of my interlocutors raised. Contrasting the sloth in operationalizing RECs in practice with the rapid societal response to the COVID-19 pandemic, she wondered if Portugal really wanted to become carbon neutral so urgently.

★★★

Seven years of researching the governance of solar energy in Portugal during this remarkable period of sectoral flux has left me with a deep sense of ambivalence about its transition. Given the urgent need of climate change mitigation, the advent of rapid solar is laudable, and credit is due to the government for steering a courageous course and playing its cards right in helping utility-scale solar come of age despite economic constraints. Yet, the implementation of these massive land-intensive installations leaves a lot to be desired, as it rides roughshod over long-standing socioecological values. The rapid, irreversible erasure of nature and culture is concomitant with these large-scale developments. In this wider perspective, it is unfortunately the antithesis of an antidote to the polycrisis we face (Koasidis et al, 2023), which requires a deeper way of valuing nature. Beyond technical fixes, solar and other development must address the climate crisis, biodiversity crisis and social inequality crisis in more conjoint than oppositional ways. Large-scale solar deployment as prioritized and practised in Portugal fails on the second and third counts. It represents a cynical framing of an energy transition, as not in the least transformative; a prolongation of top-down control of the energy sector as a money-maker for large companies prioritized by governments over other concerns like environmental or social protection.

Community-scale solar, for all the guts and persistence shown by its champions, whom I feel both admiration and gratitude for, has been successfully stalled for too long, whether with malintent by protectionist incumbency politics or by miscalculation, to really come into its own in the 2020s, given prevalent market trends. It offers lessons to other countries not as far along the curve as Portugal, to deprioritize spatially concentrated project promotion, and to offer strategic support and mandates to ensure spatially distributed solar deployment.

This should be accompanied by distribution grid strengthening and expansion, along with digitalization for two-way flows and monitoring, as well as making patient capital available to enable individual and collective investment to grow small-scale solar generation assets. Ministries that recognize the long game of highly visible payoffs in terms of an enriched and empowered population will reap the political rewards of their foresight and courage. That is a missed opportunity the Portuguese government can only look back upon regretfully, despite the continued run of leadership with the same main party at the helm since 2015. But it can – and should – invest public resources in supporting efforts to develop RECs. There are skilled and talented people with a clear vision and goodwill, eager to use this tool to benefit the Portuguese people. To keep withholding support would be political folly.

Where do I stand when it comes to balancing hope with realism and pragmatism? Somewhere in between the politics of prefiguration and negotiation. If we take the empirically most likely reading that large solar wins the day, we make this even likelier. It is analytically consistent with the empirical trend. But that leaves little scope to push for just solar energy transitions. That, after all, is the ambition, not some target of 20.4 GW by 2030. To maintain this ambition means embracing the kernel of possibility that RECs can become a major force for a different and better modality of solar development. It means critiquing the worst excesses of utility-scale solar projects, and pushing to institute a fuller accounting of socioecological impacts in how these projects are approved and implemented. This can only happen by linking meaningful financial sanctions with fulfilment of adequate standards, and not backing down with fear that foreign investors will withhold capital. Investments that water down the socioecological richness of a country serve it poorly; it will show with time.

I have cultivated a great affection for the Portuguese people and their beautiful country. I cherish the hope that this great solar adventure they have embarked on will shape up better in the next seven years towards 2030 than it did during 2017–2023. I am deeply appreciative of the openness, trust and generosity that so many interlocutors have reposed in me over the years, to help me make sense of these complex and momentous developments. I wish the story I have to tell had a happier ending, and I sincerely hope that it will lead to a better solar energy future – one that many of those featured in these pages will doubtless continue to work for, and one that I wish for my work to support and enable.

For now, I am left with platitudes like 'the future is what we make of it'. Perhaps a middling path is the pragmatic compromise. And yet, as the sun rises in Portugal, it will not usher in a just solar energy transition.

Notes

one Introduction

[1] B. Santos, 'Portugal's January-April PV installations hit 118 MW', *PV Magazine*, 21 June 2023. https://www.pv-magazine.com/2023/06/21/portugals-january-april-pv-installations-hit-118-mw/

[2] B. Santos, 'Portugal's first solar energy yields results', *PV Magazine*, 7 October 2022. https://www.pv-magazine.com/2022/10/07/portugals-first-solar-energy-community-yields-results/

[3] L. Neves, 'Brazil hits 30 GW milestone', *PV Magazine*, 2 June 2023. https://www.pv-magazine.com/2023/06/02/brazil-hits-30-gw-milestone/

two Methodology

[1] An additional work in press deserves mention: Sareen and Martin (2024).

[2] Like Portugal, these are primarily in Europe (Austria, Estonia, France, Greece, Hungary, Italy, Norway, Romania, Spain, Sweden, the Netherlands), but some also beyond (India, the United States, Uganda, Kenya, South Sudan).

three Solar Portugal 2017

[1] D. Thomas, 'Five years on ... remembering the fires of October 2017', *Portugal Resident*, 13 October 2022. https://www.portugalresident.com/five-years-on-remembering-the-fires-of-october-2017/

[2] S. Jones, 'Sun, surf and low rents: Why Lisbon could be the next tech capital', *The Guardian*, 29 October 2016. https://www.theguardian.com/world/2016/oct/29/lisbon-web-summit-sun-surf-cheap-rents-tech-capital

[3] P.S. Molina, 'Solar module factory to resume production in Portugal', *PV Magazine*, 24 May 2021. https://www.pv-magazine.com/2021/05/24/solar-module-factory-to-resume-production-in-portugal/

four Solar Portugal 2018–2019

[1] M. Bruxo, 'Monchique fire 2018: EDP and employee formally accused of negligent arson', *Portugal Resident*, 6 November 2020. https://www.portugalresident.com/monchique-fire-2018-edp-and-employee-formally-accused-of-negligent-arson/

[2] S. Sareen, 'The desirable future of solar energy in Portugal', *ECO123*, 2019. https://eco123.info/en/news/the-desirable-future-of-solar-energy-in-portugal/

[3] Reuters, 'Storm Leslie hits Portugal, leaves thousands without power', *Reuters*, 14 October 2018. https://www.reuters.com/article/uk-storm-leslie-portugal-idUKKCN1MO0BT

[4] J. Costa, 'Campanha Linha Vermelha: Red Line Campaign', *350*, 9 June 2021. https://stories.350.org/campanha-linha-vermelha-red-line-campaign/

[5] Whereas public resistance to Portugal entering fossil fuel extraction so late in the day during a climate emergency put paid to Galp Energia's consortium-building efforts back in 2018, in 2023 they partnered with TotalEnergies for joint exploration towards a 10 GW offshore wind plan instead, a sign of the changing tides. See https://www.galp.com/corp/en/media/press-releases/press-release/id/1477/galp-and-totalenergies-to-jointly-explore-offshore-wind-opportunities-in-portugal

[6] The matter of what taxes solar plants generate for municipalities and others is more complex than this, with discussion of a shift from 1.5 per cent of initial investment to 1.5 per cent of taxable income, similar to a 2 per cent revenue sharing practice for wind plants. For a solar tax overview, see: Macedo Vitorino, 'Taxation of solar plants in Portugal', 2021. https://www.macedovitorino.com/xms/files/20211103-Taxation_of_Solar_Plants.pdf

five Solar Portugal 2020–2021

[1] For content from the ENGAGER training school of April 2021, see https://www.engager-energy.net/trainingschool2/

[2] For details, see the Energy Poverty Advisory Hub website: https://energy-poverty.ec.europa.eu/index_en

[3] For an overview of the Via Algarviana, see https://viaalgarviana.org/en/

[4] The public website from this initiative is available here: https://www.observatorio-fotovoltaico.pt

[5] Details of the exhibition are on the EDP website: https://www.edp.com/en/changing-tomorrow-now/exhibition

six Solar Portugal 2022

[1] The Eurosolar4All project website is https://sunforall.eu
[2] The Smile Sintra project website is https://www.smile-sintra.com/
[3] An overview of CER Telheiras is on the Viver Telheiras website: https://vivertelheiras.pt/certelheiras/
[4] For REC legislation on collective self-consumption, see details on the ERSE website: https://www.erse.pt/atividade/regulamentos-eletricidade/autoconsumo/
[5] Given Coopérnico's frontrunner role in the 2010s in Portugal, these challenges are documented in considerable research – including mine – and grey literature, so I do not enter into more detail here.

seven Solar Portugal 2023

[1] For more details about the network, see: https://www.cost.eu/actions/CA19126/

eight Conclusion

[1] The Tamera website offers details about the eco-community and its work: https://www.tamera.org
[2] A dedicated website for Solis offers details in Portuguese: https://www.solis-lisboa.pt

References

Anderson, B., Grove, K., Rickards, L. and Kearnes, M., 2020. Slow emergencies: Temporality and the racialized biopolitics of emergency governance. *Progress in Human Geography*, 44(4), pp 621–639. https://doi.org/10.1177/0309132519849263

Batel, S. and Küpers, S., 2023. Politicizing hydroelectric power plants in Portugal: Spatio-temporal injustices and psychosocial impacts of renewable energy colonialism in the Global North. *Globalizations*, 20(6), pp 887–906. https://doi.org/10.1080/14747 731.2022.2070110

Bento, N. and Fontes, M., 2015. Spatial diffusion and the formation of a technological innovation system in the receiving country: The case of wind energy in Portugal. *Environmental Innovation and Societal Transitions*, 15, pp 158–179. https://doi.org/10.1016/ j.eist.2014.10.003

Biegel, B., Westenholz, M., Hansen, L.H., Stoustrup, J., Andersen, P. and Harbo, S., 2014. Integration of flexible consumers in the ancillary service markets. *Energy*, 67, pp 479–489. https://doi. org/10.1016/j.energy.2014.01.073

Blondeel, M., Van de Graaf, T. and Haesebrouck, T., 2020. Moving beyond coal: Exploring and explaining the powering past coal alliance. *Energy Research & Social Science*, 59, 101304. https://doi. org/10.1016/j.erss.2019.101304

Blühdorn, I., 2007. Sustaining the unsustainable: Symbolic politics and the politics of simulation. *Environmental Politics*, 16(2), pp 251–275. https://doi.org10.1080/09644010701211759

Bose, A.S. and Sarkar, S., 2019. India's e-reverse auctions (2017–2018) for allocating renewable energy capacity: An evaluation. *Renewable and Sustainable Energy Reviews*, 112, pp 762–774. https://doi.org/10.1016/j.rser.2019.06.025

Bouzarovski, S., Petrova, S. and Sarlamanov, R., 2012. Energy poverty policies in the EU: A critical perspective. *Energy Policy*, 49, pp 76–82. https://doi.org/10.1016/j.enpol.2012.01.033

Chase, J., 2019. *Solar power finance without the jargon*. Singapore: World Scientific.

Christophers, B., 2022. Taking renewables to market: Prospects for the after-subsidy energy transition: The 2021 Antipode RGS-IBG lecture. *Antipode*, 54(5), pp 1519–1544. https://doi.org/10.1111/anti.12847

Delicado, A., Figueiredo, E. and Silva, L., 2016. Community perceptions of renewable energies in Portugal: Impacts on environment, landscape and local development. *Energy Research & Social Science*, 13, pp 84–93. https://doi.org/10.1016/j.erss.2015.12.007

Fell, M.J., Pagel, L., Chen, C.F., Goldberg, M.H., Herberz, M., Huebner, G.M., Sareen, S. and Hahnel, U.J., 2020. Validity of energy social research during and after COVID-19: Challenges, considerations, and responses. *Energy Research & Social Science*, 68, 101646. https://doi.org/10.1016/j.erss.2020.101646

Fernandes, J.M. and Magalhães, P.C., 2020. The 2019 Portuguese general elections. *West European Politics*, 43(4), pp 1038–1050. https://doi.org/10.1080/01402382.2019.1702301

Ferreira, P., Araújo, M. and O'Kelly, M.E.J., 2007. An overview of the Portuguese electricity market. *Energy Policy*, 35(3), pp 1967–1977. https://doi.org/10.1016/j.enpol.2006.06.003

Gato, M.A., 2016. Adding value to urban spaces: Two examples from Lisbon. *Etnološka Tribina*, 39(46), pp 126–138. https://doi.org/10.15378/1848-9540.2016.39.04

Geels, F.W., Sareen, S., Hook, A. and Sovacool, B.K., 2021. Navigating implementation dilemmas in technology-forcing policies: A comparative analysis of accelerated smart meter diffusion in the Netherlands, UK, Norway, and Portugal (2000–2019). *Research Policy*, 50(7), 104272. https://doi.org/10.1016/j.respol.2021.104272

Gil, L. and Bernardo, J., 2020. An approach to energy and climate issues aiming at carbon neutrality. *Renewable Energy Focus*, 33, pp 37–42. https://doi.org/10.1016/j.ref.2020.03.003

Gouveia, J.P., Seixas, J. and Long, G., 2018. Mining households' energy data to disclose fuel poverty: Lessons for Southern Europe. *Journal of Cleaner Production*, 178, pp 534–550. https://doi.org/10.1016/j.jclepro.2018.01.021

Hatton, B., 2012. *The Portuguese*. Lisbon: Clube do Autor.

Hewitt, R.J., Bradley, N., Baggio Compagnucci, A., Barlagne, C., Ceglarz, A., Cremades, R., McKeen, M., Otto, I.M. and Slee, B., 2019. Social innovation in community energy in Europe: A review of the evidence. *Frontiers in Energy Research*, 7, 31. https://doi.org/10.3389/fenrg.2019.00031

Horta, A., Gouveia, J.P., Schmidt, L., Sousa, J.C., Palma, P. and Simões, S., 2019. Energy poverty in Portugal: Combining vulnerability mapping with household interviews. *Energy and Buildings*, 203, 109423. https://doi.org/10.1016/j.enbuild.2019.109423

Hsu, A., Tan, J., Ng, Y.M., Toh, W., Vanda, R. and Goyal, N., 2020. Performance determinants show European cities are delivering on climate mitigation. *Nature Climate Change*, 10(11), pp 1015–1022. https://doi.org/10.1038/s41558-020-0879-9

Jiglau, G., Bouzarovski, S., Dubois, U., Feenstra, M., Gouveia, J.P., Grossmann, K., Guyet, R., Herrero, S.T., Hesselman, M., Robic, S. and Sareen, S., 2023. Looking back to look forward: Reflections from networked research on energy poverty. *iScience*, 26, 106083. https://doi.org/10.1016/j.isci.2023.106083

Jover, J. and Cocola-Gant, A., 2023. The political economy of housing investment in the short-term rental market: Insights from urban Portugal. *Antipode*, 55(1), pp 134–155. https://doi.org/10.1111/anti.12881

Kiesecker, J.M. and Naugle, D.E., eds, 2017. *Energy sprawl solutions: Balancing global development and conservation*. Washington, DC: Island Press.

Koasidis, K., Nikas, A. and Doukas, H., 2023. Why integrated assessment models alone are insufficient to navigate us through the polycrisis. *One Earth*, 6(3), pp 205–209. https://doi.org/10.1016/j.oneear.2023.02.009

Li, T., 2007. Practices of assemblage and community forest management. *Economy and Society*, 36(2), pp 263–293. https://doi.org/10.1080/03085140701254308

Lowitzsch, J., Hoicka, C.E. and van Tulder, F.J., 2020. Renewable energy communities under the 2019 European Clean Energy Package: Governance model for the energy clusters of the future? *Renewable and Sustainable Energy Reviews*, 122, 109489.

Mafalda Matos, A., Delgado, J.M. and Guimarães, A.S., 2022. Linking energy poverty with thermal building regulations and energy efficiency policies in Portugal. *Energies*, 15(1), 329. https://doi.org/10.3390/en15010329

Magalhães, P.C., 2017. The elections of the great recession in Portugal: Performance voting under a blurred responsibility for the economy. In P.C. Magalhães (ed) *Financial crisis, austerity, and electoral politics* (pp 66–88). Abingdon: Routledge.

Magone, J.M., 2014. *Politics in contemporary Portugal: Democracy evolving*. Boulder: Lynne Rienner Publishers.

Mahoney, K., Gouveia, J.P., Lopes, R. and Sareen, S., 2022. Clean, green and the unseen: The CompeSA framework. Assessing competing sustainability agendas in carbon neutrality policy pathways. *Global Transitions*, 4, pp 45–57. https://doi.org/10.1016/j.glt.2022.10.004

McCauley, D., Pettigrew, K.A., Todd, I. and Milchram, C., 2023. Leaders and laggards in the pursuit of an EU just transition. *Ecological Economics*, 205, 107699. https://doi.org/10.1016/j.ecolecon.2022.107699

Mulvaney, D., 2019. *Solar power: Innovation, sustainability, and environmental justice*. Oakland: University of California Press.

Newell, P., 2021. *Power shift: The global political economy of energy transitions*. Cambridge: Cambridge University Press.

Palma, P., Gouveia, J.P. and Simoes, S.G., 2019. Mapping the energy performance gap of dwelling stock at high-resolution scale: Implications for thermal comfort in Portuguese households. *Energy and Buildings*, 190, pp 246–261. https://doi.org/10.1016/j.enbuild.2019.03.002

Peña, I., Azevedo, I.L. and Ferreira, L.A.F.M., 2017. Lessons from wind policy in Portugal. *Energy Policy*, 103, pp 193–202. https://doi.org/10.1016/j.enpol.2016.11.033

Pereirinha, J.A. and Pereira, E., 2023. Living wages in Portugal: In search of dignity in a polarised labour market. *Social Policy & Administration*, 57(4), pp 481–496. https://doi.org/10.1111/spol.12887

Pessoa, F., 2006. *A little larger than the entire universe: Selected poems.* London: Penguin.

Ribeiro, A.T., 2022. National recovery and resilience plan: Portugal. *Italian Labour Law e-Journal*, 15(1S). https://doi.org/10.6092/issn.1561-8048/15671

Rigo, P.D., Siluk, J.C.M., Lacerda, D.P. and Spellmeier, J.P., 2022. Competitive business model of photovoltaic solar energy installers in Brazil. *Renewable Energy*, 181, pp 39–50. https://doi.org/10.1016/j.renene.2021.09.031

Rodríguez-Pose, A., 2018. The revenge of the places that don't matter (and what to do about it). *Cambridge Journal of Regions, Economy and Society*, 11(1), pp 189–209. https://doi.org/10.1093/cjres/rsx024

Rommetveit, K., Ballo, I.F. and Sareen, S., 2021. Extracting users: Regimes of engagement in Norwegian smart electricity transition. *Science, Technology, & Human Values*, 01622439211052867. https://doi.org/10.1177/01622439211052867

Russo, M.A., Carvalho, D., Martins, N. and Monteiro, A., 2023. Future perspectives for wind and solar electricity production under high-resolution climate change scenarios. *Journal of Cleaner Production*, 404, 136997. https://doi.org/10.1016/j.jclepro.2023.136997

Santos, N. and Moreira, C.O., 2021. Uncertainty and expectations in Portugal's tourism activities: Impacts of COVID-19. *Research in Globalization*, 3, 100071. https://doi.org/10.1016/j.resglo.2021.100071

Sareen, S., 2020. *Enabling sustainable energy transitions: Practices of legitimation and accountable governance.* Cham: Palgrave Macmillan. https://doi.org/10.1007/978-3-030-26891-6

Sareen, S., 2023. Solar spectacles: Why Lisbon's solar projects matter for energy transformation. In H. Haarstad, J. Grandin, K. Kjærås and E. Johnson (eds) *Haste: The slow politics of climate urgency.* London: UCL Press, pp 234–242.

Sareen, S. and Haarstad, H., 2018. Bridging socio-technical and justice aspects of sustainable energy transitions. *Applied Energy*, 228, pp 624–632. https://doi.org/10.1016/j.apenergy.2018.06.104

Sareen, S. and Haarstad, H., 2021. Decision-making and scalar biases in solar photovoltaics roll-out. *Current Opinion in Environmental Sustainability*, 51, pp 24–29. https://doi.org/10.1016/j.cosust.2021.01.008

Sareen, S. and Martin, A. (eds), 2024. *Geographies of solar energy transitions: Conflicts, controversies and cognate aspects.* London: UCL Press.

Sareen, S., Girard, B., Lindkvist, M., Sveinsdóttir, A., Kristiansen, S., Laterza, V., Aguilar-Støen, M. and Langhelle, O., 2023a. Enabling a just energy transition through solidarity in research. *Energy Research & Social Science*, 101, 103143. https://doi.org/10.1016/j.erss.2023.103143

Sareen, S., Smith, A., Gantioler, S., Balest, J., Brisbois, M.C., Tomasi, S., Sovacool, B., Torres Contreras, G.A., DellaValle, N. and Haarstad, H., 2023b. Social implications of energy infrastructure digitalisation and decarbonisation. *Buildings and Cities*, 4(1), pp 612–628. https://doi.org/10.5334/bc.292

Sareen, S., Sorman, A.H., Stock, R., Mahoney, K. and Girard, B., 2023c. Solidaric solarities: Governance principles for transforming solar power relations. *Progress in Environmental Geography*, 2(3), pp 143–165. https://doi.org/10.1177/27539687231190656

Sequera, J. and Nofre, J., 2020. Touristification, transnational gentrification and urban change in Lisbon: The neighbourhood of Alfama. *Urban Studies*, 57(15), pp 3169–3189. https://doi.org/10.1177/0042098019883734

Silva, J.C.M. and Pereira, J.A., 2019. EDP: Portugal's main energy producer that everyone loved to hate. *The CASE Journal*, 15(6), pp 545–574. https://doi.org/10.1108/TCJ-05-2019-0050

Silva, P., Carmo, M., Rio, J. and Novo, I., 2023. Changes in the seasonality of fire activity and fire weather in Portugal: Is the wildfire season really longer? *Meteorology*, 2(1), pp 74–86. https://doi.org/10.3390/meteorology2010006

Stirling, A., 2019. How deep is incumbency? A 'configuring fields' approach to redistributing and reorienting power in socio-material change. *Energy Research & Social Science*, 58, 101239. https://doi.org/10.1016/j.erss.2019.101239

Tavares, A.O., Barros, J.L., Freire, P., Santos, P.P., Perdiz, L. and Fortunato, A.B., 2021. A coastal flooding database from 1980 to 2018 for the continental Portuguese coastal zone. *Applied Geography*, 135, 102534. https://doi.org/10.1016/j.apgeog.2021.102534

Tejada, R., 2003. Venture capital policy review: Portugal. *OECD Science, Technology and Industry Working Papers*, 2003/19. Paris: OECD. https://doi.org/10.1787/613767247743

Trahan, R.T. and Hess, D.J., 2021. Who controls electricity transitions? Digitization, decarbonization, and local power organizations. *Energy Research & Social Science*, 80, 102219. https://doi.org/10.1016/j.erss.2021.102219

Vallera, A.M., Nunes, P.M. and Brito, M.C., 2021. Why we need battery swapping technology. *Energy Policy*, 157, 112481. https://doi.org/10.1016/j.enpol.2021.112481

Wittmayer, J.M., Campos, I., Avelino, F., Brown, D., Doračić, B., Fraaije, M. et al, 2022. Thinking, doing, organising: Prefiguring just and sustainable energy systems via collective prosumer ecosystems in Europe. *Energy Research & Social Science*, 86, 102425. https://doi.org/10.1016/j.erss.2021.102425

Index

References to endnotes show both the page number
and the note number (231n3).